The Ecofeminist Storyteller

Renée Mickelburgh

The Ecofeminist Storyteller

Environmental Communication through Women's
Digital Garden Stories

Renée Mickelburgh
School of Communication and Arts
University of Queensland
St Lucia, QLD, Australia

ISBN 978-3-031-59241-6   ISBN 978-3-031-59242-3   (eBook)
https://doi.org/10.1007/978-3-031-59242-3

Cover pattern © Melisa Hasan

This Palgrave Macmillan imprint is published by the registered company Springer Nature Switzerland AG.
The registered company address is: Gewerbestrasse 11, 6330 Cham, Switzerland

Paper in this product is recyclable.

For Anthony, Evie, Jack, and Jeremy. My loves. My garden.

ACKNOWLEDGEMENTS

Writing is a social act, and this book would not have been possible without the support, encouragement, feedback and space provided by many. I must first acknowledge the traditional custodians of the lands on which I live and work: the lands of the Turrbal and Jagera people in Meanjin-Brisbane, Australia. I pay my respects to elders past, present and emerging. You are the first and greatest story*tellers* of this land we call Australia.

Family often comes last in acknowledgements but my family have been living with some iteration of this garden book for years and so it is only appropriate that I acknowledge them first. Anthony, Evie, Jack and Jeremy, thank you for your patience and love and for always being there in my search for words. To my mum, sister, father, and Carol, thank you always for lending an ear, or offers of babysitting. I love you all.

This book emerged from a PhD thesis which was written and developed during very challenging times, not least a global pandemic. I wish to thank my supervisors: Professor Elizabeth Mackinlay and Drs Natalie Collie and Melissa Harper for their kind and thoughtful words about my research and my desire to depart radically in my academic writing. Professor Mackinlay in particular laid the groundwork for this approach with her innovative, inspiring and inclusive programme: *Departing Radically in Academic Writing*. It was there I met an interdisciplinary group of scholars committed to the act of writing creatively for critical purposes. I also wish to thank the wonderful women behind the digital publications I studied: Robyn Rosenfeldt (*Pip*); Caroline Kemp (*Urban Food Street*); Georgina Reid (*The Planthunter*, now known as Wonderground). Thank you for sharing your work with the world, and making gardens an important part of glocal environmental communication.

Female friendship is extremely important to me and old friends are like diamonds: Katie and Nadine, thank you for always being there. For always being only a phone call away. You are, and always will be, my 'faraway, nearby', as the great artist Georgia O'Keefe would say.

I wish to thank the University of Queensland for providing support, both financial and in-kind. My PhD research, which this book is based on, would not have been possible without the assistance of a UQ Research Training Program Scholarship.

CONTENTS

About the Author

Renée Mickelburgh is a feminist communications scholar and Associate Lecturer in Strategic Communication based at the School of Communication and Arts, the University of Queensland, in Brisbane, Australia. At the heart of Renée's work is a strong commitment to listening and empowering the communities and organisations she works and researches with. She is interdisciplinary and collaborative in focus and embraces innovative and creative methods of communication, particularly community-based, participatory approaches, and digital methods of engagement. Renée lives and works on the lands of the Turrbal and Jagera people in Meanjin-Brisbane.

LIST OF FIGURES

Introduction

Hearing the Word Garden

What do you hear when I write the word garden?
Feminist writer-scholar Sara Ahmed says the word feminism fills her with hope
and energy and so too does the word garden for me (2017, 1).
Gardens grow in a container
but are difficult to contain
and so, they are hard to explain.
Gardens are places
where paths roam
on a collision course with memories
and ideologies
and discoveries.
Gardens are complicated dreams
filled with seeds and stories that seem
and don't seem
to be
ordinary
everyday
sweaty
homework.
Gardens are a deep acquaintance with place
in some space
and time.
Gardens are a variegated
concentrated

systematic
review
of what happens in a shared habitat
of species
and history
and forgetting on purpose.
They are a place where so many fail to admit
who decided
to share, or not to share, this place
in the first place.
Gardens are
a land use claim
within a dispossession frame
work.
The word garden is filled with soil and promise.
Something always grows in a garden
sometimes blooms
sometimes weeds.
Gardens: How to love the earth.
When gardens are a lens through which to view the world
what stories are told?

Letters of Note: The Communication of Correspondence

Letters sit quietly in university libraries, waiting for the chance to run away with the imaginations of writers and scholars. I ran away with some of these letters, the ones that contained secret matters of the garden and the world beyond its boundaries. The writer and scholar Hélène Cixous is an expert on writing about secret matters, insisting that "as soon as one writes to exhume, one secretes secrets" (2008, 178). In trying to secrete the secrets of communication, I turned to the digital world, that not-so-secret, secret place of garden stories for answers. The secret stories that had settled amongst the pages of Australian women's letters about their gardens took root in my mind as I pondered an alternate route to better communication. Australian environmental historian Katie Holmes (2011) led me to these letters in her chapter about the poet's garden, a study of the letters between poet Judith Wright and wildflower illustrator Kathleen McArthur. Holmes reveals the ways, in corresponding about their gardens, the pair developed a deep friendship that bloomed into a broader conservation ethic and action. Their almost 50-year correspondence started in 1951, when Wright lived at Mount Tamborine on the Gold Coast hinterland and McArthur resided at Caloundra, on the Sunshine Coast. The lands they lived on are the ones I too am on a first-name basis with, and so the stories about women who went before me, communicated something to the world about land as a place worth preserving, sparked my interest.

The letter-writing, activist friendship of Wright and McArthur affected me greatly (Mickelburgh, 2020). I walk, swim and fish regularly in the part of the world the pair worked so hard to preserve. It is indeed a place of pristine beauty, unmarked by the over-logged and over-loved country on its boundaries. But my connection to these letters extended further. Wright and McArthur were mothers living unconventional lives for the time: conservative, 1950s Queensland. They bonded in outrage at unfettered progress, development, and what they believed was the almost certain destruction of the Great Barrier Reef, K'Gari (Fraser Island) and the Cooloola region. The letters founded a friendship that bloomed into a determined and resolute environmental activism. Wright and McArthur breached their garden boundaries through their letters and friendship. In 1962 they helped to start the Wildlife Preservation Society of Queensland, working to protect some of Queensland's most pristine environments. They successfully fought and won major campaigns: stopping sandmining on K'Gari, preventing mining on the Great Barrier Reef, and establishing those locations and Cooloola as national parks (Wright & Snibson, 2014). Wright and McArthur were not women criticising policy from afar. They were not looking down on people or places they were far removed from. They were embedded in place, making a deep acquaintance with place, as much as the plants and people they wrote and painted. The friendship between Wright and McArthur was a charged connection; their willfulness was a spark (Ahmed, 2017, 82). Writer Margaret Somerville also ran with the Wright-McArthur letters revealing how the pair "moved quickly from the shared sense of landscape history to develop a new landscape language of gardens and wildflowers" (2004, 60). This new language of place, says Somerville, formed the basis of the pair's environmental protection work (2004, 61). It was language that emerged in the garden.

It is almost 60 years since Wright and McArthur put pen to paper and wondered, "[H]ow do we live and garden in this land?" (Holmes, 2011, 224). I found myself asking the same question, with the additional question: how do we communicate a better way of being on this land? In her personal memoir of landscape and Wright's life, Fiona Capp notes that the poet's move from "private poet to public activist" first registered for concern in her own backyard and, then, "it was just that her sense of what constituted her backyard had expanded" (2010, 224). This concept also helps answer Margaret Somerville's question about Kathleen McArthur: "what happens when a woman moves out of her garden into a wilderness that is both physical and political?" (2004, 66). What happened was an environmental movement based on story and friendship that lives on to this day. Tracey Bunda and Louise Phillips write of five principles for storying social movements: the way it "nourishes thought body and soul";

"claims voice in the silenced margins"; become "embodied relational meaning-making"; "intersects the past and the present as living oral archives"; and "enacts collective ownership and authorship" (2023, 6).

As I re-turned to the Wright-McArthur letters of friendship and gardening, I sensed an entanglement between friendship, communication and activism. Their work was "glocal". In *Feminist Ecologies: Changing Environments in the Anthropocene* (2018), editors Lara Stevens, Peta Tait and Denise Varney describe this concept of "glocal". They say:

Whether it involves maintaining a garden for food, involvement in an anti-pollution campaign, or marching to raise awareness about rape, the most powerful global responses to ecofeminist issues today are happening at a local level—at the level of the glocal (2018, 7).

I followed Wright and McArthur's footprints around—Cixous might call them rootprints (Cixous & Calle-Gruber, 1997)—and saw how they ran parallel with Sara Ahmed's concept of willfulness. Through their art and poetry, and their letters, the pair seemed to be searching for a language of the land, one that could communicate its beauty and value to the rest of the human world. As I viewed their "willfulness archive", that is those "documents that are passed down in which willfulness comes up, as a trait, as a character trait" (Ahmed, 2014, 13) I wondered how gardening willfulness was being represented through contemporary stories. Ahmed says willfulness is a "style of politics when we are not willing to go with the flow" (2017, 82) and it is true here too for Wright and McArthur. The pair paid a heavy price for not going with the flow. These gardeners were "labelled 'ratbags' and suspected of communist sympathies, if not direct links with the KGB" while also being insulted by men "who did not expect to discuss anything on the ecology of a coral reef, coastal management, biological diversity etcetera, with a woman" (Holmes, 2011, 250). Yet, their willfulness stopped the flow of unfettered development; the flow of mining; and the flow of money for ecologically destructive projects. They kept things flowing; their willful flow of words and art between each other and the world has kept the ocean flowing through the reef, and communication about environmental concerns flowing through the years. The use of correspondence as a spark for their environmental communication, built on a foundation of artistic sensibilities, signalled a way forward. This book is the result of my search for today's willful women; those women telling stories of their garden, and also saying something about contemporary environmental communication.

PART I

Letter One: (Summer)[1]

Dear, Kathleen,

I have come inside to write after a day in my garden. Today's writing is a day when the word garden means nothing is growing very well and everything is in half. I write of stunted zucchinis, of weather that is too hot and of rains that are avoiding me while grey soil crumbles in my hand. I write about the way the flowers are scrawny and will struggle against the day's angry heat that we expect, unexpectedly, at this time of year (Mickelburgh, 2020a, 147). Do you remember when drought decimated your garden too? Kathleen, my garden, like yours, is also where the word home grows. It is home school, homework, work from home, at home, on my home. Margaret Somerville described your ethic as "based on home, the local, the embodied and (the) direct, daily interactions with landscape" (2004, 2). Something significant is going on in the home, it's just that now, more people know about it. Your home sounds like mine.

My garden is place of everyday action with global consequences. My garden is fenced in, built up, watered down (when the rains come). My garden is the place where secrets lay in wait, just below the soil. I think of my garden as one of many gardens in Australia, rather than an Australian garden. Is that what you thought of too when you spoke of gardens and place? Somerville describes the letters between you and Judith Wright as a "shared language of landscape" (2). I too now find myself working to

---

[1] Sections of this series of letters appear in 'Writing strange letters in the garden, with love and future' Swamphen: A Journal of Cultural Ecology.

discover, or perhaps rediscover, that language. I fear a sense of urgency on this research path; I worry it is a language becoming lost or already forgotten. You both wrote about gardens and flowers, but in doing so you also made notations about things happening outside your garden boundaries and borders. Your correspondence breached garden containment lines, traveling across backyards and through time and space. I too am wondering how today's garden stories are suspended forever in an on-line, web of time where they become entangled in a past, present and future time-line.

Your dear friend Judith's daughter Meredith McKinney collaborated with Patricia Clarke to collate a volume of Wright's letters called "With Love & Fury". The title was a nod to a letter Wright wrote to her cousin's husband, Peter Warwick (Wright, 2006, 237). Warwick was a public relations officer for the Country Party and Judith wrote him a strongly worded letter about that party's position on the rights of Indigenous peoples. Your friend believed in the power of the written word, not just through her poetry but in her communication. She would write up to 12 letters a day to friends and colleagues, as well as organisations and government leaders. She rarely kept her own letters, but you retained much of the correspondence from your friend. You knew it was to be treasured.

And so, I sign off as Judith once did.

With Love and Fury

XXXX

CHAPTER 1

Communication on Common Ground: An Alternative Paradigm

The world is in trouble—ecological devastation, climate change and oppression along gender, race and class lines pose significant and catastrophic threats. This book's focus is humble in comparison; it follows French feminist philosopher Hélène Cixous' advice for writing, in that it must wander near to what is close to the earth—humus (2008, 3). This book takes this humble approach literally. In considering environmental communication; that its emergence from an individual place—the home garden—to a collective space—the digital realm—provides for an essential, communicative paradigm. Specifically, it is a search sparked by women concerned with the intersection of the local and the global. But it is also a search for more. The great American writer Ursula K. Le Guin says human communication cannot be reduced to information because "the message not only involves, it is, a relationship between speaker and hearer" (2004, 187). I sensed something about relationships going on within the digital stories of the home gardens I saw and heard and read on the screen. All the stories I studied represented a range of first-person narratives. Such narratives, says ecofeminist Karen Warren, are important to feminism and environmental ethics because of the way they emphasize the importance of being "'in relationship' with others, including the non-human environment" (1996, 27). This sense of relationship, became the search for what leading Australian ecofeminist philosopher Val Plumwood described as an "alternative communicative paradigm" of the world perceived in "communicative and narrative terms" (2002, 203). Such a paradigm must

R. Mickelburgh, *The Ecofeminist Storyteller*, https://doi.org/10.1007/978-3-031-59242-3_1

tell stories in ways that show "a deep and loving acquaintaince" (p. 203). Plumwood was a philosopher, not a communications scholar, and yet she saw the value of communication as a method for change. And it is that insight that sparked my curiosity. Plumwood's words became a lens through which to view a very specific, contemporary, and increasing, style of communication.

Ecofeminism as a theory and practice emerged in the 1970s, as a way of critiquing the dual oppressions, and their connections, of gender and nature. Understanding ecofeminism as a method of communication, however, is much more recent, and emerging process (Singer, 2020, 1).The imperative for ecofeminist communication research, however, is growing more urgent. To grapple with ecological devastation, not to mention gender oppression, is a huge amd unweildy undertaking. To garden, in comparison, is a relatively humble activity. To grapple with both the garden and ecological devastation is to care. It is important to remember that to garden, is also, to love. Member of the Citizen Potawatomi Nation and plant scientist Robin Wall Kimmerer makes the provocative claim that to love a garden is also to have the garden love you back (2013, 123). Similarly, Plumwood notes the importance of a "deep and loving acquaintance" with place. With those guiding intensions I use the garden as the centrepoint to this book, to search for a new, or perhaps nourished, form of ecofeminist communication. In doing so, this book asks simple questions of a simple place: the garden. Specifically, what does a garden do to help women storytellers into being and becoming more sensitive to the world around them? How might contemporary Australian women's stories that emerge from the garden place, and take root in the online space, be a method of environmental communication? What happens when that communication is digital in nature and storied in structure, and is rooted in themes of community, connection, care, and compassion? I try to answer these questions by examining the digital stories of Australian women's gardening lives.

Communication aims for connection and eco-domesticity and environmentalism is blooming in blogs and websites (Padilla Carroll, 2016). This growth in glocal communication is not without its critics, and often justifiably so: there are always weeds among the flowers. So-called *everyday environmentalism* has been criticised for its "individualization of responsibility" whereby powerful political structures and institutions avoid interrogation and clear action to minimize harm in favour of an emphasis on smaller actions like planting a tree or recyling in order to save the world (Maniates,

2001, 33). Feminists too have critiqued the individualist model for the continued (unpaid) responsibility it lumps on women, a responsibility to once again clean up the mess and take charge of caring responsibilities. While writing this book I worried about the risk of perpetuating and glorifying the historical legacy of more unrecognised and unpaid care work (MacGregor, 2004, 2006). These are important and legitimate concerns which exemplify affect theorist Lauren Berlant's "collective norm of obligation" that exists within stories which communicate an enthusiasm for a return to eco-domesticity (2004, 41). Additionally, there are critiques about the politics of contemporary, networked communication media. Political theorist Jodi Dean calls it "communicative capitalism" (2009). While users may be entangled in networks of "enjoyment, production and surveillance" communicative capitalism is an "economic-ideological form wherein reflexivity captures creativity and resistance so as to enrich the few as it placates and diverts the many" (2010, 11). Dean has devoted a whole book to the subject of blogging as a lens through which to theorise the political implications of networked communications, considering the disagreement between those who espouse the "democratic ideals of access, inclusion, discussion and participation" and those who point to the shift in power relations, and the convergence of wealth among the few (2010, 11). Perhaps the truth is somewhere in the middle. I can't help but remain hopeful, remembering Cixous' words that "all the world is a garden" (2008, 150). If all the world is a garden, and if the garden is also the place that we love and might love us back, the place we make a deep and loving acquaintance with, then perhaps the garden is also a place for place better understanding the world and ways we communicate in it. Analysing stories that grow from the garden for their communicative potential (or lack thereof) does not absolve powerful social and political forces from responsibility. Perhaps it might do just the opposite.

Australian Indigenous writer Melissa Lucashenko says, "it is taken for granted that the landscape that has fed and nurtured our ancestors has shaped us in deep unspoken ways. Who understands this? Gardeners, perhaps" (2003, 17). I am not Aboriginal or Torres Strait Islander, and make no connection to Indigenous ways of knowing and being, but I am a gardener with a deep and loving acquaintance of the land I dig and grow and weed and water. In doing so, I also know that gardens are a place of connection and stories. Therefore, this study is not quantitative in kind. There is no number crunching, or big data to contend with. Instead, this is a qualitative analysis guided by ecofeminist principles, which are in essence, both subjective and reflective. By their very nature, the stories I study may constitute communicative capitalism, but they also represent a digital memory-making. That is, they are a mediation between "self and society" (Poletti, 2020, 17). They are creating cultural and mediated memories (Van Djick, 2004, 2007), whereby individual accounts, produced through media technologies are "instruments and objects of inscription and communication" (2007, 39). These personal inscriptions are operating within a digitally intimate moment, where technology mediates connections between the private and the public (Dobson et al., 2018).

Cultivating my understanding about this eco-communicative form has required wandering through my own metaphorical gardens and the gardens of others. The stories I study are rooted in the digital realm. I have sought out three digital plots created by Australian women gardeners. I have found a garden to listen to; a garden to look at; and a garden to read. There is precedent in this respect. Other media scholars like Jose van Djick (2007) have also followed this path, examining stories of the self (blogging), aural recordings and camera footage to consider the way the individual also becomes the collective. And so, I have listened to *Pip* podcasts, where Australian women living permaculture lives tell garden stories of care and compassion. To consider the ways these stories upheld emotional narratives, and the complex ways a compassionate ecofeminist care ethic became entangled amongst them, I have interpreted this soundwork as communicative groundwork. In my second case study, I wandered from this soundwork to sightwork, watching and re-viewing photographs that depicted the growth and sudden decline of a much-loved neighbourhood verge garden: *Urban Food Street*. Doing so, I considered the precarity and promise of what it means to break down borders and boundaries and reclaim the commons in a world of private property ownership. Finally, I wandered towards the digital garden wordwork of *The Planthunter*,[1] collecting from that website a series of personal-public digital diaries detailing one woman's journey to a new life and garden in an isolated river cottage. This wordwork left me contemplating blurred boundaries between the human and more than human and what happens when what is contained becomes uncontained, through writing and digital dissemination. Examining this soundwork and sightwork and wordwork is a search for communicative groundwork. My investigative method has required

---

[1] Since this analysis was conducted The Planthunter has re-named itself and broadened its scope. It is now known as Wonderground.

clutching the words of feminist materialist scholar Karan Barad's concept of "re-turning" in my hands. That is, turning something over and over and breathing new life into it (2014, 168). In doing so, this book considers how these stories *re*-tell, *re*-view and *re*-write the human and nonhuman world through the lens of gardens. In essence, how they re-communicate the natural world.

ECOFEMINIST WRITING AS A METHOD OF INQUIRY

Ecofeminism continues to evolve, and similarly the ground I write on is not static. A garden is an experiment had in place, and so in keeping with the garden metaphor I position this book as a garden in which I also experiment experiment with words in a scholarly world. I experiment with an inside-outside, backwards-to-go-forwards writing style. I write in mimicry, but also to embody the poetic-prose legacy of feminist scholars and writers that have blazed the trail for me to follow. Ecofeminist Susan Griffin planted this seed of my style when I went wandering with her book *Women and nature* (1978). To explain the relationship between language and women and nature Griffin wrote groundbreaking prose which was "like poetry, and like poetry always begins with feeling" (1978, xv). Griffin encourages an embrace of creativity as part of a method of inquiry, writing, "if before I failed to see that a fear of nature and a fear of creativity must be inextricably associated, it is only because, schooled in my own culture and its paradigms, I have failed to see that nature and creativity share the same origin, are born in the same breath. (Griffin, 1982, 646). Part of Griffin's creative exposé lies in the powerful way she juxtaposes the historical evolution of science with the simultaneous burning of witches (Bullis, 2015, 125). Griffin's writing with feeling is embedded with critical thought and detailed research, carving a theoretical, but also methodological trail. Her words scrape away at the undergrowth, clearing weeds and making space for another way of ecofeminist writing and knowing, but also another way of communicating. Griffin's writing is a way of story*telling* which doesn't fit the genres of a documentary or an essay, rather it is "fragmented, metaphorical, and discontinuous" (Cantrell, 1994, 226). Her language unsettles the voices that resist the voices of women and nature. To justify her approach, Griffin writes that to understand "the logic of civilised man"; she went underneath that logic, "writing associatively, and thus enlisting my intuition or uncivilised self" (1978, xv). Her writing style shattered the illusion that language justifies the domination of nature, but also illustrates the relationship between science and patriarchy and the domination of nature and women.

A creative paying attention allows one to take notice of what comes near. There are many different ways to pay attention, and creative processes have long been used to explain phenomena that can not be tabled or quantified. I attempt to do this by writing from the ground by "paying attention to what's going on and 'staying open to what's in (my) vicinity'" (Berlant & Stewart, 2019, 34). Feminist scholar Elizabeth Mackinlay writes of the way she has paused at the work of French feminist philosopher Simone Weil. It is a pause that prompted her to pay attention to words and the way they might strengthen our capacities for living and loving response-ably (2022). Elizabeth St Pierre describes this as the work of post-qualitative inquiry as that place where "something in the world that is unintelligible and unrecognisable within existing categories and practices 'kicks back', sticks, and takes hold" (2018, 604). The stories I analyse here chose me; they stuck to me and took hold rather than me choosing them. They came near and touched me, as Ahmed might say. Gardens for me are a place of happiness and as Ahmed notes "happiness puts us into intimate contact with things" (2010, 23). These stories might be considered Ahmed's "good 'hap'", that Middle English word "suggesting chance" (22). I chose these stories, which were all published or aired between 2017 and 2019, by good hap but they also chose me.

My writing often weaves through, up and around different theories and stories. It is writing that wanders, but it is not aimless: it is writing as a method of inquiry (after Richardson, 1997, 2017). To write "as a method of inquiry" (Richardson, 1997, 87) is to do the method of unearthing, exposing faded paths and ways of doing and being and revealing a more desired ecofeminist way to communicate environmental problems—and solutions—through. I am not the first to write as a method of inquiry. While feminist and post-structuralist scholars paved the desire line towards a creative approach to academic inquiry, Laurel Richardson followed suit with her *Fields of Play* (1997). As part of my inquiry I engaged with a group of scholars committed to *Departing Radically in Academic Writing* (Mackinlay & Madden, 2024) as a way to privilege alternative and creative methods of inquiry. And so, mimicking the great science fiction writer Ursula K. Le Guin I carry all these writing methods in my carrier bag (Le Guin, 1989, 65).

This book is a write-ful response to my questions about gardens and communication and being more sensitive to the world around us. I am writing about the small stories of gardens to consider ways digital communication of some of these Australian places might reweave the world (an original ecofeminist goal), this approach seemed not only the most appropriate but the most hopeful. It allowed me to consider the ways digital stories that emerge from home gardens represent the "glocal"; that is "how the global relies on the local, that is, on place" (Stevens et al., 2018, 7). Some say metaphor reveals "the existence of determinate but nonsingular meanings" (Ceccarelli, 1998, 399). Considering the garden as not only a local place, but part of a global environmental endeavour made the use of metaphor essential. With it comes the understanding that these multiple meanings empower both the writer and the reader. They reduce the author's hegemonic control "and add to the insurgent power of the audience"; they provide "strategic ambiguity" to generate support from conflicting groups; while also offering the potential for how an audience "might" respond to a text (Ceccarelli, 1998). The destruction and degradation of our environment, the multiple and conflicting interests involved require such conflicting support for change to occur.

The Australian Garden: A Brief History in Time

Before I take you into the gardens I have studied, it is important to understand the foundations of the field. Australia is a land colonised, with a long history of violence and disposession and denial against First Nations people. The history of the Australian land the garden sits on must be considered within all this pondering. How do you communicate a loving acquaintance with a land that has been stolen? Can it even be done? Here I am guided by Lucashenko again who points to small stories, or "earth-speaking", which "might help us find a way through" (2003, 17). Yet, you may ask, why are there no Indigenous Australian women's stories here? Why were these particular garden stories included and not others? I too am worried by this absence, and I know I will worry about it long after this book is published.

I feel it is important to highlight, rather than hide, this lack. I will start my explanation guided by the words of bell hooks, who warns against the *privileged* writing *about*[2] the marginalised or disenfranchised as an act of political solidarity. Such action, she warns, risks being "as much an act of colonizing appropriation as the more apparent conventional modes of

---

[2] Italics are author's own.

white supremacist capitalist patriarchal dominance" (1999, 42). I did feel compelled to seek out Australian Indigenous women's voices in the digital garden world, a place in and of itself a space of white privilege. To my dismay and disappointment, I failed in that endeavour. My inability to source such stories does not mean Indigenous Australian women are not telling or producing stories about gardens. Rather, the small number of digital stories I came across at the time my study was carried out, hints at the continuation of the historical legacy of erasure of Aboriginal women since colonisation, and an invisibility of the important garden work that Indigenous Australian women have long carried out. The internet is a vast space with algorithms that engage, but also erect barriers around our understanding and our connection with others and their knowledge. There is no doubt that Aboriginal and Torres Strait Islander women are gardening, and telling stories about those gardens. But where, when, how and why are still questions that hang silent around this study.

A secret can stay a secret by simply not speaking of it. Absence, just as much as silence, keeps things a secret. Colonisation is the not-so-secret secret embedded in the Australian garden soil. In Australia, the garden, and the gardener, plants her seeds in earth that has been stolen. The garden is a British "idea" that colonised Australia. It is the space where "a stolen land became claimed and 'owned' through (a) central 'ritual of habitation'" (Holmes, 1999, 152). Aileen Moreton-Robinson, a Goenpul woman of the Quandamooka people and Australia's first Indigenous Distinguished Professor, might call this colonial secret one of "white possessive logics" (2015, xii). Such a secret, she says, occurs where any sense of "belonging, home and place enjoyed by the non-Indigenous subject—colonizer/migrant—is based on the dispossession of the original owners of the land and the denial of our rights under international customary law" (3). For settler-descendants like me, exploring this harsh situatedness is often difficult, but it is imperative. There are uncomfortable memories buried in a garden. Often, they have been there long before a gardener arrives to water the plants and tend the seedlings. Anthropologist Deborah Bird Rose says harm often occurs "at a distance from us, or in contexts that we are not well trained to see and understand" (2015, 129). A garden is not a distant place. Harm has happened *here*, not out *there*. The Australian garden *is* the place where damage has been done. Gardeners might make a space beautiful, but harm has occurred, but often stayed a secret, in order for a gardener to be there at all. Pulling back the leaves and wandering the secret path through the garden leads to uncomfortable places, where uncomfortable things happen. What I wonder is if it is also a place of repair?

Absence needs to be interrogated, and silence illuminated. Anti-racism must also involve critiquing limits of representations white people make. Aileen Moreton-Robinson's work in *Talkin' Up to the White Women*, is clear. She insists that the subject position "middle class white women" must be made visible in white feminist academic discourse in order to reveal the power effects of its invisibility and the complexity and limits to relations between white feminists and Indigenous women (2000, 184). Failing to make whiteness visible "leaves whiteness uninterrogated, centred and invisible" (184). I take seriously Moreton-Robinson's statement that "to change the power relations between these two groups of women is more complex than giving voice, making space or being inclusive within a white feminist politics of difference" (p. 184). In response, rather than analyse the work of Australian Indigenous women, I have carried the concerns of Indigenous and black female academics with me on my journey, reflecting often how I might make visible and interrogate the whiteness and power issues of settler-colonial women's stories I study.[3] I hope the knowledge produced from my analysis might do as Moreton-Robinson urges us to: contribute to informing feminist theory and used in anti-racist practice (2000, 184). I take seriously the words of Tuck and Wang, that "decolonization is not a metaphor" and avoid considering gardens as a metaphor for decolonisation in an effort to acknowledge the ongoing the way this might risk "settler moves to innocence" (2012, 1). That is, gardens and the way they are communicated as important and may contain vital informaiton about environmental communication. But they should not be used as an vasion tactic that attempts to water down settler guilt.

PLOT 1: A GARDEN TO LISTEN TO

Garden stories grow like weeds and sunflowers, existing wherever there is light, some soil and humans to cultivate. So how to choose which ones to study? Before I tell you more about *why* I write the way I do, I must tell you *what* I have written of. I pressed my ear close to a series of the Australian-based *Pip* Magazine (n.d.) podcast episodes, which aired

---

[3] While one of the stories I studied was not that of a white woman, she was, as Moreton-Robinson points out, a migrant and all migrants "share the benefits" of Indigenous dispossession (2015, 17).

between 2017 and 2018, to consider what happens when we listen stories about and around gardens. These podcasts provide the matter through which my methodological digging begins. Podcasts are relatively new forms of digital story-*telling*. If telling *is* listening, as Ursula K. Le Guin insists (2004), that is, it is creates connection between story-*teller* and story-*listener*, then podcasts provide the ideal medium through which to better understand what is exactly is being connected when some women speak about their garden, and what that connection might do. This telling is listening matter echoes like birdsong in my head each time I ponder the stories in front of me. I use it as a prompt, to ask questions about women and the home and the more than human and what it means to live a good life. Le Guin says, "listening is not a reaction, it is a connection" (2004, 196) and so I have listened closely to the telling for the ways this connection works. I search for the ways these stories connect women to their gardens; the ways their stories connect them to their audience and to me and also, perhaps, how these stories connect them to the wider world. Podcasts embody Le Guin's description of a conversation. They are a shared event, that is "the listener's listening enables the speaker's speaking. It is a shared event, intersubjective: the listener and speaker entrain with each other" (Le Guin, 2004, 198).

I stumbled across *Pip* podcasts when I was wondering what women say when their work wanders close to home, when it becomes a re-turn-to-home environmentalism. Permaculture, an Australian-developed philosophy, emphasises care for people and the earth, and reduced consumption. In and of itself Permaculture is not specifically a garden practice, but the garden is an essential and integral part of permaculture. The women in the *Pip* podcasts link permaculture closely to everyday environmentalism through permaculture's three core tenets: "earth care, people care and fair share" (Rosenfeldt, 2017, 4:24). It seemed logical, then, to follow the way care and compassion flow from gardens into communicative methods. I wondered what I might hear when women tell stories of care and compassion in the home garden space. Those two words—care and compassion—are words that have a history of cornering women, containing them, put them in the home corner. Does *re*-turning them to a private, gendered place simply make the environment another mess for women to clean up? There are many questions on my lips as these women's words fill my ears: what happens when garden stories of compassion mobilise an ethic of care contingent on a re-turn to the home environmentalism? What happens when these stories of care signal an alternative way of life, an alternative good life? How are the emotions of pleasure and pain caught up in these compassionate spaces? Does re-turn to home environmentalism, therefore, only reinforce Berlant's prediction about the search for the good life proving cruelly optimistic (2011), or does it say something else?

PLOT 2: A GARDEN TO LOOK AT

Searching through images on a screen about a different, unconventional way to use a piece of land means thinking deeply about the complexities of use. It means considering the history of use, the way stories travel across place and time, and the way these images are used in the online space. I went back in time to 2016, when *Urban Food Street* (n.d.) published images of its verge garden on Facebook and wondered about the way those images delighted the visual senses with a socially-mediated story about food grown on the verges of a suburban neighbourhood in southeast Queensland. What matters in a garden story that signals a reclaiming of the common space? Understanding the commons is understanding many forms of form. The creation of urban gardens, says feminist scholar Silvia Federici, is one of the most important (2019, 105).

I turn to Karan Barad's theory of diffraction, or more specifically her concept of "re-turning" to consider the *Urban Food Street* pictures (2014, 168). Reclaiming a verge space for the growing of neighbourhood food plants seeds that spark a remembering, this is an often an unknown memory of a time when the garden place was common space. When a visual story tells the tale of a contemporary reclaiming of the commons, a suburban street-scape becomes an imagined world and the viewer becomes a time traveller. Even the way I write of this plot, mostly in past tense, reveals the precarity of gardens: *Urban Food Street* is a physical entity no longer exists. Yet ecofeminists have long advocated for re-turning to what was in order to imagine what could be (Mies, 2014; Mies & Bennholdt-Thomsen, 1999, 2001) and more recently contemporary scholars of all persuasions have echoed their sentiments (Gibson, 2018; Johnston, 2006, 2008). *Urban Food Street* pictures a contemporary place and time, but it also recalls another time, when women tried to reclaim their common land after British and European enclosure, and the establishment of private property (McDonagh, 2019; McDonagh & Griffin, 2016). Re-claiming, however, also risks re-turning to a silence and it is here I am troubled with a stories that remain silent about the history of claiming and re-claiming stolen Australian land. I have re-turned to Berlant as a way through this part of the path I think along. Berlant warns that the commons concept should not be displayed as a utopian solution but rather it should be a "a powerful vehicle for troubling, troubled times" (2016, 395). To use a garden as an example of common space means thinking the uses of use—who has the privilege of use and who does that use extend to?

PLOT 3: A GARDEN TO READ

Questions of communication and connection become entangled once more when I read the words of another Australian gardener who writes her garden in the digital space. I followed Georgina Reid, editor of *The Planthunter* ("Planthunter," n.d.), as she settled into a new life and garden by the side of a river. The diary, once the preserve of the private, is made public in this online space. Writing and publishing her diaries and memoirs between 2017 and 2019 Reid's words are the only case study of the three that focuses on an inherently personal narrative, rather than collective. What does a diary do and say when it is written for one, but also for many? *The Planthunter* describes itself as "weaving a rich story of connection and respect between humans, culture, plants and environment" and in Reid's digital diaries I discover the way old connections meld into a new narrative. I wonder what role connection plays in this communication.

My interest in the world of *The Planthunter* morphed into an exploration of the way, through Reid's own writing, plants, people, and place intersect, overlap, and become entangled through the garden space. In the garden and through her digital memoirs, Reid grapples with issues of boundaries and borders; of relationships between humans and non-human; and the complexities of time. She blurs her garden boundaries with her actions and thoughts, which in turn become digital fragments of a larger story. Her narrative intrigued me with the way it made me think about the "deep acquaintance with some place" (Plumwood, 2002, 231) but also the to consider what happens when autobiography is mediated in the digital realm (Poletti, 2020). If, as Ursula K. Le Guin suggests, "reading is a means of listening" (2004, 209) then what do Reid's readers learn by becoming in communion with what is on her mind? Connections build through this reading-writing relationship and words and ideas re-animate the matter of place.

REFERENCES

Ahmed, S. (2010). *The promise of happiness*. Duke University Press.

Ahmed, S. (2014). *Willful subjects*. Duke University Press.

Barad, K. (2014). Diffracting diffraction: Cutting together-apart. *Parallax: Diffracted Worlds—Diffractive Readings: Onto-Epistemologies and the Critical Humanities, 20*(3), 168–187. https://doi.org/10.1080/13534645.2014.927623

Berlant, L. (2004). Introduction: Compassion (and withholding). In L. Berlant (Ed.), *Compassion: The culture and politics of an emotion* (pp. 1–14). Routledge.

Berlant, L. (2011). *Cruel optimism*. Duke University Press.

Berlant, L. (2016). The commons: Infrastructures for troubling times. *Environment and Planning D: Society and Space, 34*(3), 393–419. https://doi.org/10.1177/0263775816645989

Berlant, L., & Bojarska, K. (2019). The hundreds, observation, encounter, atmosphere, and world-making. *Journal of Visual Culture, 18*(3), 289–304. https://doi.org/10.1177/1470412919875404

Berlant, L., & Stewart, K. (2019). *The hundreds*. Duke University Press.

Bird Rose, D. (2015). Dialogue. In D. Bird Rose, R. Fincher & K. Gibson (Eds.), *Manifesto for living in the anthropocene* (pp. 127–131). Punctum Books. https://doi.org/10.21983/P3.0100.1.00

Bochner, A. P., & Ellis, C. (1992). Personal narrative as a social approach to interpersonal communication. *Communication Theory, 2*(2), 165–172. https://doi.org/10.1111/j.1468-2885.1992.tb00036.x

Bullis, C. (2015). Retalking environmental discourses from a feminist perspective: The radical potential of ecofeminism. In J. G. Cantrill & C. L. Oravec (Eds.), *The symbolic earth: Discourse and our creation of the environment* (pp. 123–148). University Press of Kentucky.

Cantrell, C. (1994). Women and language in Susan Griffin's woman and nature: The roaring inside her. *Hypatia, 9*(3), 225–238. https://doi.org/10.1111/j.1527-2001.1994.tb00459

Ceccarelli, L. (1998). Polysemy: Multiple meanings in rhetorical criticism. *The Quarterly Journal of Speech, 84*(4), 395–415. https://doi.org/10.1080/00335639809384229

Cixous, H. (2008). *White ink: Interviews on sex, text and politics*. Acumen.

Cixous, H., & Calle-Gruber, H. (1997). *Rootprints: Memory and life writing*. Routledge.

Dean, J. (2009). *Democracy and other neoliberal fantasies communicative capitalism and left politics*. Duke University Press.

Dean, J. (2010). *Blog theory: Feedback and capture in circuits of drive*. Polity Press.

Dobson, A. S., Carah, N., & Robards, B. (2018). Digital intimate publics and social media: Towards theorising public lives on private platforms. In A. S. Dobson, B. Robards, & N. Carah (Eds.), *Digital intimate publics and social media* (pp. 3–27). Palgrave Macmillan.

Federici, S. (2019). *Re-enchanting the world: Feminism and the politics of the commons*. PM Press.

Gibson, K. (2018). *Introduction: Food as urban commons and community economics*. University of Western Australia Publishing.

Griffin, S. (1978). *Woman and nature: The roaring inside her*. Harper & Row.

Griffin, S. (1982). The way of all ideology. *Signs, 7*(3), 641–660. https://doi.org/10.1086/493904

Griffin, S. (1993). Red shoes. In R.-E. B. Joeres & E. Mittman (Eds.), *The politics of the essay: Feminist perspectives* (pp. 1–11). Indiana University Press.

Holmes, K. (1999). Gardens. *Journal of Australian Studies, 23*, 152–162. https://doi.org/10.1080/14443059909387485

hooks, b. (1997). *Wounds of passion: A writing life*. Henry Holt.

hooks, b. (1999). *Remembered rapture: The writer at work*. Henry Holt.

Johnston, J. (2006). Who cares about the commons? In M. A. Gismondi, J. Goodman, & J. Johnston (Eds.), *Nature's revenge: Reclaiming sustainability in an age of corporate globalization* (pp. 39–72). Broadview Press.

Johnston, J. (2008). Counterhegemony or bourgeois piggery? food politics and the case of foodshare. In Middendorf, G., & Wright, W (Eds). *The fight over food : producers, consumers, and activists challenge the global food system*. Pennsylvania State University Press.

Le Guin, U. K. (1989). *Dancing at the edge of the world: Thoughts on words, women, places*. Grove Press.

Le Guin, U. K. (2004). *The wave in the mind: Talks and essays on the writer, the reader, and the imagination* (1st ed.). Shambhala.

Lucashenko, M. (2003). Not quite white in the head. *Griffith Review, 2*, 7–15.

MacGregor, S. (2004). From care to citizenship: Calling ecofeminism back to politics. *Ethics & the Environment, 9*(1), 56–84. https://doi.org/10.2979/ETE.2004.9.1.56

MacGregor, S. (2006). *Beyond mothering earth: Ecological citizenship and the politics of care*. UBC Press.

Mackinlay, E. (2022). *Writing feminist autoethnolgraphy: In love with theory, words, and the language of women writers*. Routledge.

Mackinlay, E., & Madden, K. (Eds.). (2024). *Departing radically in academic writing: Alternative approaches to writing methods in qualitative research*. Routledge.

Maniates, M. F. (2001). Individualization: Plant a Tree, Buy a Bike, Save the World? *Global Environmental Politics, 1*(3), 31–52. https://doi.org/10.1162/152638001316881395

McDonagh, B. (2019). *Gendering protest and the commons.* Paper presented at the Proceedings of the International Conference of Historical Geographers, Warsaw, Poland.

McDonagh, B., & Griffin, C. J. (2016). Occupy! Historical geographies of property, protest and the commons, 1500–1850. *Journal of Historical Geography, 53,* 1–10. https://doi.org/10.1016/j.jhg.2016.03.002

Mies, M. (2014). No commons without a community. *Community Development Journal, 49*(SI), i106–i117. https://doi.org/10.1093/cdj/bsu007

Mies, M., & Bennholdt-Thomsen, V. (1999). *The subsistence perspective: Beyond the globalised economy.* Zed Books.

Mies, M., & Bennholdt-Thomsen, V. (2001). Defending, reclaiming and reinventing the commons. *Canadian Journal of Development Studies, 22*(4), 997–1023. https://doi.org/10.1080/02255189.2001.9669952

Moreton-Robinson, A. (2000). *Talkin' up to the white woman: Aboriginal women and feminism.* University of Queensland Press.

Moreton-Robinson, A. (2015). *The white possessive: Property, power, and indigenous sovereignty.* University of Minnesota Press.

Padilla Carroll, V. (2016). The radical possibilities of new (feminist, environmentalist) domesticity: Housewifery as an altermodernity project. *Interdisciplinary Studies in Literature and Environment, 23*(1), 51–70. https://doi.org/10.1093/isle/isw013

Pip Magazine. (n.d.). *Pip Podcasts.* https://pipmagazine.com.au/category/podcasts/

Planthunter. (n.d.). *The Planthunter.* https://theplanthunter.com.au/

Plumwood, V. (2002). *Environmental culture: The ecological crisis of reason.* Routledge.

Poletti, A. (2020). *Stories of the self: Life writing after the book.* New York University Press.

Richardson, L. (1997). *Fields of play: Constructing an academic life.* Rutgers University Press.

Richardson, L. (2017). Writing: A method of inquiry. In N. K. Denzin & Y. S. Lincoln (Eds.), *The Sage handbook of qualitative research* (5th ed., pp. 1410–1444). Sage.

Rosenfeldt, R. (Host) (2017). # 3: What is permaculture? with Hannah Moloney [Audio podcast episode] Pip Podcast. *Pip Magazine.* https://www.pipmagazine.com.au/podcasts/what-is-permaculture-hannah-moloney/

Singer, N. R. (2020). Toward Intersectional Ecofeminist Communication Studies. *Communication Theory, 30*(3), 268–289. https://doi.org/10.1093/ct/qtz023

St. Pierre, E. A. (2018). Writing post qualitative inquiry. *Qualitative Inquiry, 24*(9), 603–608. https://doi.org/10.1177/1077800417734567

Stevens, L., Tait, P., & Varney, D. (2018). Introduction: "Street-fighters and philosophers": Traversing ecofeminisms. In L. Stevens, P. Tait, & D. Varney (Eds.), *Feminist ecologies: Changing environments in the anthropocene* (pp. 1–22). Palgrave Macmillan.

Tuck, E., & Yang, K. W. (2012). Decolonization is not a metaphor. *Decolonization: Indigeneity, Education and Society, 1*(1), 1–40.

Urban Food Street. (n.d.). *Facebook: Urban Food Street.* https://www.facebook.com/urbanfoodstreet

Van Djick, J. (2004). Mediated memories: Personal cultural memory as object of cultural analysis. *Continuum: Journal of Media and Cultural Studies, 18*(2), 261–277. https://doi.org/10.1080/1030431042000215040

Van Djick, J. (2007). *Mediated memories in the digital age.* Stanford University Press.

Wall Kimmerer, R. (2013). *Braiding sweetgrass.* Penguin Books.

Warren, K. (1996). The power and the promise of ecological feminism. In K. Warren (Ed.), *Ecological feminist philosophies* (Vol. 12, pp. 125–146). Indiana University Press.

CHAPTER 2

A Matter of Method: Ecofeminist Writing as a Method of Inquiry

Ecofeminism is a concept that carves a circular path of theory, practice, and creativity (Gaard, 1993; Mies & Shiva, 2014) in its search for transformative discourses (Bullis, 2015). This chapter explores more deeply my positioning as an Australian gardener-mother-ecofeminist communications scholar and links it with the feminist braiding that underpins this book's methodology. Foundational to ecofeminism is the understanding that there is an ideological interconnection between the patriarchal domination of women and nature and other oppressed groups and ways of thinking (Plumwood, 1993; Stevens, 2018; Warren, 1996). The dualisms that position the patriarchy, reason and culture above all that is feminine, nature, and emotion are deeply embedded ways of thinking that have caused harm to those in the position of subject (Plumwood, 1993, 43). Within this dualism "the properly human (as reason, coded male) is seen as opposed to and divorced from nature (coded female), as the animal and the ecologically situated body, just as the non-human is hyper-separated from ethics and culture (Plumwood, 2018, 106). This "logic of domination" (Warren, 1996, 21) sustains and justifies the dominations of women and nature, as well as races and classes (Bullis, 2015; Warren, 1994). Plumwood describes these interconnected oppressions as a web because a web contains "both one and many, both distinct foci and strands with room for some independent movement of the parts, but a unified overall mode of operation, forming a single system" (1994, 79). As the world became entangled in global market systems this web of oppressions, "encircles the whole globe and begins to stretch out to the stars" (p. 80). What is needed, to reweave this dominating web, says Plumwood, is both practical and methodological cooperation (1993, 1994).

To write *of* communication, as well as *about* communication, is to consider the imperative for a border-less, boundary-less world of community-in-common. To do so I attempt to re-turn to, but also to re-write, an ecofeminist way of communicating. Ecofeminist writing has often (but not always) required writing outside boundaries dividing the scholarly paper from the poetic essay and acknowledging "poetic vision as a form of knowledge and as one of the important steps in the process of global transformation" (Diamond & Orenstein, 1990, viii). This writing-research-life braiding is what drew me to ecofeminism in the first place, despite its "great and difficult" presence (Diamond & Orenstein, 1990, xv). Although an emerging field, ecofeminist communication scholars say it is essential that posthumanist feminist approaches also be considered in the context of ecofeminist communication in order to decenter the focus on any one human, or more than human entity (Singer, 2021, 2). And so, as part of the argument for my critical-creative approach is that a contemporary ecofeminist method must consider the affective implications of experience and emotion. Engaging with boundary-less writing, experimenting with different ways to tell this research story, has allowed me to better understand the affectiveness in the communication I have been studying. Some of this writing-as-research might seem unconventional, some might seem confusing. But so too is the world we are living in, and the challenges we face.

Reweaving Feminist Writing Forms to Transform

Storytelling as a method of communication is in no way an invention of the global north and ecofeminism should never claim as much. Indigenous knowledges have long used braiding metaphors to tell stories and explain concepts. In Australia, where I am from and where the stories I study for this book emerge, Aboriginal and Torres Strait Islander peoples have been telling stories, orally and visually, for millennia as a way of ensuring continuous care for country. As Palyku writer Ambelin Kwaymullina writes, for generations Australian Indigenous women "gathered and ground the seeds that nourished our kin, danced the rhythms of the earth in the ceremonies by which the world is renewed, and sang the stories of our

Ancestors as we moved through Country" (2018, 195). Ngugi/Wakka Wakka writer and scholar Tracy Bunda, and co-researcher Louise Gwenneth Phillips say, "we see story as the communication of what it means to be human, that tells of emplaced, relational tragedies, challenges and joys of living" (2018, 3). For Aboriginal people, stories carry meaning "transgenerational law and life spoken across time and place" (8). Storytelling has also long been a tradition of western folklore. In Irish folklore, for example, groups would gather at each others houses in the evenings to tell stories through music, dancing and singing (Harvey, 1989).

Storytelling is transformative in the most ordinary ways. Writer Joan Didion famously said, "we tell ourselves stories in order to live" (1979, 11) while philosopher Hannah Arendt said, "storytelling reveals meaning without committing the error of defining it" (1968, 158). Storytelling is a human communication practice involving the telling of narrative to make sense of the world. It also requires imagination for the capacity to generate empathy, encourage agency, blur boundaries, and cultivate contingency (Hutchison, 2010; Jackson, 2002, 2013; Jacobs, 2002; Whitebrook, 2001). It is impossible to exclude imagination in storytelling, because "to reconstitute events in a story is no longer to live those events in passivity, but to actively rework them, both in dialogue with others and within one's own imagination" (Jackson, 2013, 34). Storytelling reveals important issues, ideas, and events (e.g., Hutchison, 2010; Winskell & Enger, 2014).

Scientist and member of the Citizen Potawatomi Nation Robin Wall Kimmerer encourages a deep acquaintance with gardens in *Braiding Sweetgrass* (2013). When questioned about the best way to restore relationship between land and people she says,

> My answer is almost always "plant a garden". It's good for the health of the earth and it's good for the health of people. A garden is a nursery for nurturing connection, the soil for cultivation of practical reverence. And its power goes far beyond the garden gate—once you develop a relationship with a little patch of earth, it becomes a seed itself. (2013, 126)

Wall Kimmerer's *Braiding Sweetgrass* (2013) breaks ground with the way it weaves together Indigenous knowledges and storytelling with scientific research and historical events to create a narrative around ecological conservation that is both unique and engaging. Her story is also a way of communicating what is, and what could be when it comes to the protection of the natural world. While I do not claim Indigenous knowing, it is important to recognise the work of Indigenous storytellers and their capacity for change, education and inspiration. I hope that I too, might follow Wall Kimmerer's lead, and others, through my attempt to re-weave linguistic stiches of the oppressive master web, showing ways of writing that might work towards the liberation of women, and the more than human, and all the other many groups in the world that face oppression and domination. For Australian feminist scholar Elizabeth Mackinlay, to tell, speak, or write story in research is to be "both troubled and troubling by the relationships it invokes" (2015, 1439).

Structurally, the telling of narrative differs from other discursive forms of text such as report writing or historical annals or chronicles (White, 1987, 44). Narratives, usually, contain an identifiable beginning, middle and end, characters, and points of view. Events are recounted through a plot and a "logic that makes recounted events meaningful" (Polletta, 2006, 99). Storytelling embraces the performative. It is not always what is written, but rather it also often involves a speaker (or writer or performer) as storyteller and listeners as audience (Langellier & Peterson, 2011). There is a temporal dimension to narrative. Events within a story can be real or imagined, but often unfold in a sequence. Everywhere, says sociologist Laurel Richardson, humans make sense of their temporal worlds through narrative (1997, 29). These narratives, others say, are "not necessarily fiction and yet also not indisputable fact, narratives construct a kind of life-flow, bestowing a pattern upon life and the happenings within it" (Hutchison, 2016, 117). It's important to recognise, however, that narratives gesture to a Euro-centric understanding of the word narrative. Bunda says the term narrative is a colonising concept that has "snuck its way in". She and Phillips suggest story focuses on all forms, whether they be "spoken, gestured, dramatized, painted, drawn, etched, sculpted, woven, stitched, filmed, written and any combination of these modes and more" (2018, 3). Story, therefore, means more than beginning, middle and end. It is about connection and relationships.

Reweaving a web of oppression is a complicated sewing process. To consider how 'glocal' communication can emerge from gardens requires bringing many different strands of thought together to consider how they might work as one. I consider the deep friendship that emerged "between the leaves" (Holmes, 2011) of the letters between poet Judith Wright and wildflower illustrator Kathleen McArthur and take Margaret Somerville's question as my own to ask, "what happens when a woman moves out of her garden into a wilderness that is both physical and political?" (2004,

66). Different threads are needed and understanding how they might come together cooperatively is an undertaking that emerges slowly. A web, reminds Plumwood, can continue to function, even with damage to some of its parts. She points out that "rarely if ever can it be said, 'once we have cut this section, solved this problem, all the rest will follow, other forms of oppression will wither away" (1994, 80). Parts can be conflicted and even move in opposing directions for a short while. Strategies for reweaving, then, require a methodological process of "cooperative movement", where cooperation encompass both theory and action and there is "a choice of strategies or of possibilities for theoretical development" (1994, 80). My attempt to reweave the threads of oppression Plumwood speaks about required a methodology that gleans writing tools to allow me to become willfully and usefully distracted by what comes near (Ahmed, 2014, 83). With these tools I sketch rough paths through and into a post-qualitative world of ecofeminist thinking and doing and being.

I am not the first to attempt to use language as a methodological garden tool, to dig up and better understand the way language oppresses, but also how it can be both connective and resistive. To write as a method of inquiry (Richardson, 2017; St. Pierre, 2018) is to become entangled in a strange post-qualitative world. I have reached for the "too strange" or "too much" that St. Pierre insists must be grasped, rather than pruned from analysis (2018, 607). Gleaning, according to Merriam Webster, can be both the gathering of leftover grain in a field, or to "gather information or material, bit by bit" (Glean, 2021). In eighteenth century England gleaning was "an inclusive right" claimed by the poor and a "mainly female occupation" (Whyte, 2011, 160). It seemed fitting then, that as a feminist communication scholar I too am engaging in a gleaning of sorts as part of my research method. I have gleaned a range of critical-creative writing methods in order to "write my way into" my too strange, too much work of post-qualitative inquiry (p. 606). My wandering methodology gleans Laurel Richardson's concept of writing as a method of inquiry (Richardson, 1997, 2017). For Richardson "writing is a process of discovery" (1997, 93) and a feminist concern because "wherever text is being produced, there is the question of what social, power and sexual relationships of production are being *reproduced*" (1997, 57). Richardson insists that we ask questions of what our writing is doing; how is it reproducing dominant systems and how might it challenge those systems? We need to think about who we are speaking and writing to, what voice we are using and what criteria.

I grasped onto writing as a method of inquiry in a bid to show ways of thinking and findings that I could not properly articulate through scholarly writing, particularly in my attempts to reveal the affective connection digital garden stories communicate. Sociologist Laural Richardson's emphasis on communication as community heavily influenced my approach (1997, 79). Richardson encourages a jumble of experimental writing, the blurring of genres and the mixing of writing methods. This might include narratives of the self, visual presentations, cybertexts, performance science and much more (1997, 92). Within my own work I was grappling to explain the *feeling* of digital intimacy and so experimented with the "poetic representation of lives" (Richardson, 1997, 139). Such writing engages with the "pause" of speaking, that is the true way we experience speech in the body. Doing so engages with the embodiment of research and writing, encouraging a response that is both cognitive and sensory (p. 143). Communication as community allows for communal participation, association, a deeper understanding of the locale, temporality, trust and empathy (1997, 79). This was what I was searching for.

Among other creative fragments seeded throughout this book, part of this writing as method of inquiry can be seen in the strange letters that punctuate its contents. My strange letters to Kathleen McArthur, a wildflower artist long passed, was a way to emphasise how this inquiry has felt "too strange" and "too much" (St. Pierre, 2018, 607). Alongside my research into women's digital stories of gardens has been the discovery that such garden writing between and among Australian women is not new, and that its power and potential to bloom into broader environmental communication and activism has an historical track record. Australian poet Judith Wright and McArthur were from the region I hail from. Their work, in poetry, prose, letters and cards, emerged from stories about their gardens and evolved into an intense communication campaign to save Cooloola, the Great Barrier Reef and K'gari (Fraser Island) from mining. The strange letters I wrote in my research journey helped me to better understand this friendship and how their writing and activism laid the groundwork the communicative methods I study today. They are research notes, of sorts, an attempt to both illustrate and understand their connection.

The other part of this attempt at understanding ecofeminist communication, is the use of *hundreds* throughout this book. That is, paragraphs in a series of a hundred words at a time, that tries to emulate the affectiveness of the garden stories I study. This is a homage to the work of affect theory scholars Kathleen Stewart and Lauren Berlant. It is my attempt to glean the feeling of un-containment communicated in their book of theoretical poems *The hundreds* (2019). The two scholars write that *the hundreds* "is an experiment with keeping up with what's going on", but which are not "events of knowing, units of anything, revelations of realness, or facts" (p. 5). In mimicry to Berlant and Stewart, and following Cixous' insistence that "what is most true is poetic" (1997, 3), I too have punctuated this book with poetic-prose in the form of theoretical hundreds. Doing so, I am attempting to "produce ideas about the ordinary of composition, not just on the page, but also in the encounter with words, worlds, people, animals, and a variety of things" (Berlant & Bojarska, 2019, 290). Writing this way is both humble and willful, in that it is writing that is close to the earth (Ahmed, 2014; Cixous, 2008, 3). To write from the ground, as Berlant and Stewart might say (2019, 34), is to search for common ground, where stories might signal communication and, perhaps, communion, in common. I write this way remembering Le Guin's warning about merging poetry and prose, that the borderline between the two "is one of those fogshrouded literary minefields where the wary explorer gets blown to bits before ever seeing anything clearly" (1989, 104). My writing as method of inquiry is not pure mimcry, it is the product of a jumble of influences. Susan Griffin's experiment with form in her essay "Red shoes": part italicised memoir, part traditional essay influences this style as I try, as she did, to ask, "is it possible to write in a form that is both immersed and distant, far seeing and swallowed?" (1993, 11). There is also no doubt that the path of experimental-critical writing has been carved by bell hooks, whose work *Wounds of passion* (1997) is impacted and influenced many other writers, of whom I count myself as one. If I, as a communication scholar am a "communicating (human) studying humans communicating", then I am also "inside what (I am) studying" (Bochner & Ellis, 1992, 165). A garden is an experiment with some place. It is, Robin Wall Kimmerer insists, "both a material and spiritual undertaking (2013, 123). In grappling with all that a garden is and could be—a material, emotional, spiritual, memorable, and now digital place—my writing, too, became an experiment had in a place.

My work is grounded in the dirt of Australian soil and stories. But just as other feminist scholars like Mackinlay have done before me, I have been swept up in the possibilities diffractive narrative methodology holds for unsettling colonial ways of doing business" (2016, 178) and have held the prefix "re" close while wandering and wondering about this space. Clutching Barad's concept of a re-turning in my hands—that is, turning something over and over and breathing new life into it (2014, 168)—I have considered how these stories *re*-tell, *re*-view and *re*-write the human and non-human world through the lens of gardens. To garden and write of that garden is the diffractive theory in motion of Karen Barad's "re-turning". Like Barad I found myself re-turning to words and histories as part of my method, cutting them together apart; aerating them; breathing new life into old ideas and ways of being (2014, 168). For Barad this way of re-turning is not simply returning, that is it is not "reflecting or going back to a past that was, but re-turning as in turn it over and over again" (p. 168). Barad describes this way of learning as diffraction. Barad uses the earthworms to explain this re-turning methodology. Re-turning is, she says, "turning the soil over and over—ingesting and excreting it, tunnelling through it, burrowing, all means of aerating the soil, allowing oxygen in, opening it up and breathing new life into it" (2014, 168). To use a diffractive methodology, she says, "is a critical practice for making a difference in the world" (Barad, 2007, 91). To consider the way garden stories re-turn us to communication is to consider the ways garden stories re-turn to concepts of the good life; to discussions about pleasure and pain; and to the ecofeminist concerns about consumption, caring and compassion work. It is also re-turning to the matter of garden stories as stories of common ground. To study the stories of gardens is to regularly re-trace stories of garden lives. It is viewing these stories in the sense that they are "diffracting", there is no clear separation between the then and now, but rather "matter itself is diffracted, dispersed, thread through with materializing and sedimented effects of iterative reconfigurings of spacetimematterings, traces of what might yet [have] happen[ed]" (2014, 168). Thinking this way is considering the way stories help us understand "actual and existing worlds" (Mackinlay, 2016, 172). For Mackinlay, engaging a re-turning to story in the Baradian sense is to engage with threads that twist and turn between bodies as matter and the potential for re-imagining difference (2016, 173). If, as Cixous says, the whole world is a garden, then we need to start re-imagining ways for our garden to grow.

References

Ahmed, S. (2014). *Willful subjects*. Duke University Press.

Arendt, H. (1968). *Men in dark times*. Harcourt, Brace and World.

Barad, K. (2007). *Meeting the universe halfway: Quantum physics and the entanglement of matter and meaning*. Duke University Press.

Barad, K. (2014). Diffracting diffraction: Cutting together-apart. *Parallax: Diffracted Worlds—Diffractive Readings: Onto-Epistemologies and the Critical Humanities, 20*(3), 168–187. https://doi.org/10.1080/13534645.2014.927623

Berlant, L., & Bojarska, K. (2019). The hundreds, observation, encounter, atmosphere, and world-making. *Journal of Visual Culture, 18*(3), 289–304. https://doi.org/10.1177/1470412919875404

Berlant, L., & Stewart, K. (2019). *The hundreds*. Duke University Press.

Bochner, A. P., & Ellis, C. (1992). Personal narrative as a social approach to interpersonal communication. *Communication Theory, 2*(2), 165–172. https://doi.org/10.1111/j.1468-2885.1992.tb00036.x

Bullis, C. (2015). Retalking environmental discourses from a feminist perspective: The radical potential of ecofeminism. In J. G. Cantrill & C. L. Oravec (Eds.), *The symbolic earth: Discourse and our creation of the environment* (pp. 123–148). University Press of Kentucky.

Bunda, T. (2018). Seeing the Aboriginal sovereign warrior woman. *Lifted Brow, 40*, 4–5.

Cixous, H. (2008). *White ink: Interviews on sex, text and politics.* Acumen.

Cixous, H., & Calle-Gruber, H. (1997). *Rootprints: Memory and life writing.* Routledge.

Diamond, I., & Orenstein, G. F. (1990). *Reweaving the world: The emergence of ecofeminism.* Sierra Club Books.

Didion, J. (1979). *The white album.* Simon & Schuster.

Gaard, G. (1993). *Ecofeminism: Women, animals, nature.* Temple University Press.

Glean. (2021). *Merriam Webster Dictionary* (Vol. 2020). https://www.merriam-webster.com/dictionary/glean

Griffin, S. (1993). Red shoes. In R.-E. B. Joeres & E. Mittman (Eds.), *The politics of the essay: Feminist perspectives* (pp. 1–11). Indiana University Press.

Harvey, C. B. (1989). Some Irish women storytellers and reflections on the role of women in the storytelling tradition. *Western Folklore, 48*(2), 109–128. https://doi.org/10.2307/1499685

Holmes, K. (2011). *Between the leaves: Stories of Australian women, writing and gardens.* UWA Publishing.

hooks, b. (1997). *Wounds of passion: A writing life.* Henry Holt.

Hutchison, E. (2010). Unsettling stories: Jeanette Winterson and the cultivation of political contingency. *Global Society, 24*(3), 351–368.

Hutchison, E. (2016). *Affective communities in world politics: Collective emotions after trauma.* Cambridge University Press.

Jackson, M. (2002). *The politics of storytelling: Violence, transgression, and intersubjectivity.* Museum Tusculanum Press.

Jackson, M. (2013). *The politics of storytelling: Variations on a theme by Hannah Arendt* (2nd ed.). Museum Musculanum Press.

Jacobs, R. N. (2002). The Narrative Integration of Personal and Collective Identity in Social Movements. In *Narrative Impact* (1st ed., pp. 205–228). Routledge. https://doi.org/10.4324/9781410606648-11

Kwaymullina, A. (2018). You are on Indigenous land: Ecofeminism, Indigenous peoples and land justice. In L. Stevens, P. Tait, & D. Varney (Eds.), *Feminist ecologies: Changing environments in the anthropocene* (pp. 193–208). Palgrave Macmillan.

Langellier, K. M., & Peterson, E. E. (2011). *Storytelling in daily life: Performing narrative*. Temple University Press.

Mackinlay, E. (2015). Making an appearance on the shelves of the room we call research: Autoethnography-as-storyline-as-interpretation in education. In P. Smeyers (Ed.), *International handbook of interpretation in educational research* (pp. 1437–1456). Springer.

Mackinlay, E. (2016). *Teaching and learning like a feminist storying our experiences in higher education*. Sense Publishers.

Mies, M., & Shiva, V. (2014). *Ecofeminism*. Zed Books.

Phillips, L. G., & Bunda, T. (2018). *Research through, with and as storying*. Routledge.

Plumwood, V. (1993). *Feminism and the mastery of nature*. Routledge.

Plumwood, V. (1994). The ecopolitics debate and the politics of nature. In K. Warren (Ed.), *Ecological feminism* (pp. 64–87). Routledge.

Plumwood, V. (2018). Ecofeminist analysis and the culture of ecological denial. In L. Stevens, P. Tait, & D. Varney (Eds.), *Feminist ecologies: Changing environments in the anthropocene* (pp. 97–112). Palgrave Macmillan.

Polletta, F. (2006). *It was like a fever: Storytelling in protest and politics*. University of Chicago Press.

Richardson, L. (1997). *Fields of play: Constructing an academic life*. Rutgers University Press.

Richardson, L. (2017). Writing: A method of inquiry. In N. K. Denzin & Y. S. Lincoln (Eds.), *The Sage handbook of qualitative research* (5th ed., pp. 1410–1444). Sage.

Singer, N. R. (2021). Toward Intersectional Ecofeminist Communication Studies (vol 30, pg 268, 2020). *Communication Theory, 31*(4), 1022–1022. https://doi.org/10.1093/ct/qtaa020

Somerville, M. (2004). *Wildflowering: The life and places of Kathleen McArthur*. University of Queensland Press.

St. Pierre, E. A. (2018). Writing post qualitative inquiry. *Qualitative Inquiry, 24*(9), 603–608. https://doi.org/10.1177/1077800417734567

Stevens, L. (2018). From the female eunuch to White beech: Germaine Greer and ecological feminism. In L. Stevens, P. Tait, & D. Varney (Eds.), *Feminist ecologies: Changing environments in the anthropocene* (pp. 115–133). Palgrave Macmillan.

Wall Kimmerer, R. (2013). *Braiding sweetgrass.* Penguin Books.

Warren, K. (1994). Introduction. In K. Warren (Ed.), *Ecological feminism* (pp. 1–7). Routledge.

Warren, K. (1996). The power and the promise of ecological feminism. In K. Warren (Ed.), *Ecological feminist philosophies* (Vol. 12, pp. 125–146). Indiana University Press.

White, H. V. (1987). *The content of the form: Narrative discourse and historical representation.* Johns Hopkins University Press.

Whitebrook, M. (2001). *Identity, narrative, and politics.* Routledge.

Whyte, N. (2011). Custodians of memory: Women and custom in Rural England c. 1550–1700. *Cultural and Social History, 8*(2), 153–173. https://doi.org/10.2752/147800411X12949180694263

Winskell, K., & Enger, D. (2014). Storytelling for social change. In *The handbook of development and social change* (pp. 189–206). John Wiley & Sons.

PART II

Letter Two: (Autumn)

Dear Kathleen,

When I decided to write about women and gardens, I never dreamed of the ways my work would wander from the page. I imagine the leaders in the Academy will say, we asked you to write about women's gardens and the stories they tell, what does that have to do with connection, communication, and community? I found myself riffing from Virginia Woolf to explain. When I told them I would speak about women and the stories they tell about the gardens they love, I sat at my computer for a long time to wonder what these words meant. They might mean a few words about gardens and stories and women. The title women and gardens might mean, and you may have meant it to mean, "women and what they are like"(1927/1977, 3), or it might mean women and the gardens they tell stories about; or it might mean the stories that are written about women and gardens, or it might mean all three.[1] Yet just like Woolf, I have found myself unable to find a "nugget of pure truth" (4). Promises of the good life stake a claim in the garden.

Gardens are rituals of the soil. They are also a colonial ritual linking the past with the present. Katie Holmes says these rituals manifest in myriad ways: from the way we order this environment, through to the design and materials used to create (1999, 152). They require us to construct

---

[1] The rest of this famous sentence is not a direct quote, it is a palimpsest response to Woolf's further words in *A Room of One's Own (3)* that *"it might mean women and the fiction that they write; or it might mean women and the fiction that is written about them.... Or it might mean that somehow all three are inextricably mixed together"*.

boundaries and borders to prevent the out there from getting in and the in here from getting out. Gardens stake a claim about place; in Australia they stake a claim on a stolen place. They make promises about what happens when you live a certain way. There is the promise that something will grow if you tend to it, if you care for it. Gardeners discover, especially when there is drought or flood, that promises are easily broken. Garden stories prompt the reader to ask, like Cixous did, "to feel rich, in myself, what does that mean?" (2008, 16). I've re-turned to these stories again and again, following the faint trace of a thread linking them all. To listen and watch and read garden stories one must consider the dreams and desires gardens are made of. I wonder, are these dreams compostable, or combustible?

The threat of combustion is a real and present danger now, in places not far from the garden borders. Dreams can quickly become things of nightmares: fantasies that fray (Berlant, 2011, 3). The rainforest mountains I love to wander across were razed by fire a few years ago. Are garden stories filled with compassion, or just an example of what Lauren Berlant describes as cruelly optimistic fantasies? Are they the object of my desire that has become "an obstacle to [my] flourishing" (1)? Gardens are places where life is composted, but also land filled. They are places where one gets deeply acquainted, rather than briefly associated. They are places where dreams, ideas and fantasies live in harmony with everyday actions. As I work my garden I dream of my research. As I dig up the soil I dream of the women who have dug in spaces like this before me, and those who dig now. I want to highlight the way their work is important, that they say something about paying attention, and care and compassion. That is why I've re-turned to these stories again and again, aerating them, breathing new life into them as Karen Barad might say (2014, 168). I worry I expect too much of garden stories.

With Love and Fury,
XXXX

CHAPTER 3

Pip Podcasts: When Telling Becomes Listening

A container can be anything: a sling; a pot; a bottle; a shell; a net; a box. It just needs to be something to hold and receive (Haraway, 2016, 118; Le Guin, 1989, 166). *A container, then, can also carry theory* (Le Guin, 1989, 165). *I had the perfect container sitting in the bottom of my garden shed: a harvest basket. Its potential lay in the promises it held, promises to both contain and collect. One fills a harvest basket with hopes and dreams, all the wonderful and nourishing things. It becomes an object of desire.*

I re-traced my footsteps, re-turned to the shed and re-discovered that harvest basket there, sitting right where I left it. Its emptiness sat like an accusation. It contained nothing and I needed to fill it with words that had not been written, but could still be gathered. Engulfed in dust and a complex weave of cobwebs, my harvest basket needed to be un-forgotten. Its neglect was evidence of my headlong rush towards the future (Bird Rose, 2004, 18). *It was unmoved and now, I hoped, unforgotten.*

I had a troublesome task for this harvest basket. Clasping words of care and compassion and re-turning them to this container had implications. These words were heavy and cumbersome; I worried about lugging them through weeds and bushes, getting snagged on branches. At times, it seemed,

---

Elements and themes that emerge in this chapter were developed in Mickelburgh, R. (2020) Compassion in the Garden or just more women's work? *Emotions—History, Culture, Society* 4(1), 146–166.

33

these words were slowing me down, putting me at a disadvantage. My failure to question the complex weaving of these words within the basket pointed to my lack of imagination. Or maybe I had been distracted by their promise of happiness (Ahmed, 2010).

My writing and thinking about theory resembled an imaginary dream that I had struggled to contain. I worried that the words of care and compassion I clung to would go to waste once more if I couldn't settle on a feminist story for them. My presence traced a fine line around garden stories, but it also traced a line around other worlds. Writing care and compassion troubles the past with the present in a desire to communicate the future. I slowly realised I might be weighing down the sisters of the soil. I sense, with unease, that there is trouble ahead.

Her paths are memory trails, where her compassion wanders in search of story. She knows little of worlds in other times and places, but she has "beat the bounds" of her parish paths alongside the men to secure the memory of her belonging (Whyte, 2011, 156). *Along these walking paths is where "narratives of custom and place (are) sustained and reinforced through…social meanings and memories attached to the physical landscape"* (Whyte, 2011, 154). *She is as familiar with the paths as the plants in her common garden.*

The paths she walks are where the essential "lines of communication from old to young" fold the past into the present (Whyte, 2015, 933). *This world is not just geography; it is about meaning and memory. In this common space wandering along the paths imagines the footsteps of females so they are not forgotten, even though they are often erased from the official papers of history* (Whyte, 2011). *This pathway is a conversation space. It tracks time and place, but never presumes the local is the same thing as the communal* (Berlant, 2004, 3).

The great science fiction writer Ursula K. Le Guin once wrote that"telling is listening" because "listening is not a reaction, it is a connection" (2004, 196). This chapter examines a series of Australian permaculture podcasts and considers what happens when sound becomes voice and telling becomes listening. I viewed the podcasts as containers, of sorts. They were containers that contained and un-contained garden stories. The *Pip Podcasts* I write about here aired between 2017 and 2018, and were an offshoot of the Australian written and published *Pip* Magazine, an Australian-based permaculture publication. This chapter considers the *Pip Podcasts* through the lens of "telling is listening" and searches not for reaction, but for connection (Le Guin, 2004, 196). Ursula K. Le Guin's sentiments about soundwork suggested to me that story-*telling* garden worlds might communicate the environment in a way that written or visual stories can not. In the podcasts, words are not visible, but they are audible. Words, laughter, sadness and silences are all part of the affective sound-work of podcasts. The women speaking in these podcasts used sound and voice as story-*telling*. I wondered, what connection could be found in listening to these podcasts? How does composting women's soundwork with gardening groundwork matter in the world of environmental communication? How might this mingling become a transforming? (Ahmed, 2019, 22).

Podcasts are time travellers, in that they contain stories that dissolve from the body and re-emerge withing the digital. Podcasts travel along, and across, time and space. They are often described as digital audio or visual files (although in most cases they are accessed as audio files). Both speakers, and listeners, exercise a degree of agency. Speakers can produce podcasts for relatively low cost and minimal technology (compared to large, mass or multi-media companies). Listeners can "both time-shift and place-shift their listening and viewing habits" (Haygood, 2007, 518). Many radio shows have blurred their production into the digital space (often appearing as both a radio show and a recorded podcast) and so Johnson's definition of radio applies equally to podcasts; that is, it is a personal medium "that creates a sense of community and privacy simultaneously" (2008, 98). Podcasts are convenient, accessible, often episodic, and always downloadable (Chadha et al., 2012). One subscribes to a podcast and, unlike radio, can choose where, when and even how they listen to a specific podcast (Chadha et al., 2012; Haygood, 2007). To consider what was happening with the words I listened to, I first had to consider some questions: what are podcasts and what is permaculture and what happens when they mingle? How does composting women's sound work with gardening groundwork matter in the world of environmental communication? How does this mingling become a transforming? (Ahmed, 2019, 22). My answers circled back and around to matters of time and space, affective intimacy, and digital connection.

Podcasts are more than a simple technological tool for listening to stories. They contain the potential for affective connection. Podcasts are stories that also engage a digital intimacy. That is, they connect strangers—both listeners and speakers—through voice and that voice and those stories often elicit emotion. Stacey Copeland describes this soundwork as "an intimate aural medium", which provides for a deeply affective experience for all involved (2018b, 211). Voice recordings provide an "aural junction point between the body, the self, the character of performance and listener" (2018b, 221). Copeland draws on Ahmed's work on "affective economies"

(2004a) to highlight the affective capacities of podcasts. Affectiveness operates when emotions circulate. Ahmed focuses on the "stickiness" of emotions; the way they "align individuals with communities—or bodily space with social space—through the very intensity of their attachments" (Ahmed, 2004a, 119). While Ahmed is speaking about the ways emotions stick to people, Copeland applies this affective theory to podcasting. When a listener dons headphones, Copeland says, they become "immersed in an affective discourse of human experience through listening and connecting" (2018b, 212). Media scholar Van Dijck studied recorded music rather than podcasts (and of course there are significant differences between the two), her work in this sonic space also has relevance in this context. José van Dijck has noted how "the transfer of emotive affection from the brain on to the technology and materiality of audio recordings shows ow memory acts out in the spaces between individual reminiscences and shared experiences (2007, 90). Aligning this thinking in terms of podcasts can therefore help in understanding how they are simultaneously individual and intimate, mediated and communal experiences.

There is a sense of trespass, to podcasts. Listeners can access them through multiple devices, and they can be heard publicly or privately (via headphones and mobile phones). Listeners are (mostly) silent while listening into other people's conversations, lives, and worlds. Immersion into this "simultaneously interior and exterior sonic experience" generates both intimacy and connection (Llinares et al., 2018). Le Guin writes that hearing unifies, making listening a connection, where "we don't so much as respond, as join in" (2004, 196). Podcasts don't allow listeners to speak back to those speaking to them, but the material and affective capacities of podcasts mean they are a powerful and communicative tool. They are intimate phenomena with the ability to which circulate emotions through bodily acts of speaking and listening and the technological connection they require.

Virginia Woolf famously wrote that "whenever you see a board up with 'trespassers will be prosecuted', trespass at once" (1952, 125). Podcasting then, when understood as an affective story*telling* experience, has feminist consequences because sound takes up space. When that sound is the sound of women's voices taking up space in places they had been unable to previously: women's stories trespass. Feminist scholars Raechel Tiffe and Melody Hoffman, who operated their own feminist podcast for several years, say podcasts become sites resistance when they take up this sonic space (2017). Women podcasters are embracing the sound of their own voice, rather than allowing their sound to be policed into emotionless voices commonly heard in traditional, mass media. In general, Tiffe and Hoffman say, "the more you sound like yourself the better" (2017, 117). They point to the work of sound scholar Yvon Bonenfant who says, "sound isn't solid but takes up space" because sound is "a "field of vibration that moves through matter", allowing others to hear it and feel it (2014). Podcasting, therefore, is both an affective endeavour and experience. Stacey Copeland draws on Ahmed to illustrate how podcasts are "sound-work...soaking in affect" (2018b, 212). They are "radiogenic media"; that is, the podcast medium serves as a platform "outside the scopophilic voyeuristic male gaze of popular visual mediums" (2018a, 218). Here then, is where podcasts, trespass. Radio has long been a medium that has excluded or ridiculed women's voices, shaming them for the difference in their pitch and tone.

When podcasts are seen as an affective, communicative method with the ability to trespass, they hold potential as objects of resistance and tools of environmental justice. When they contain performative, personal narratives they trespass across disciplinary boundaries (Langellier, 1989, 244) built up by big budget, male dominated multi-media companies controlled by patriarchal interests and powers. Their ability to travel far from the boundaries and borders and the time and space of their original creation remains a major strength. They can be recorded in one moment but listened to many moments later. The only access requirement is some form of everyday, digital technology to allow this aural performance to play. To borrow the words of Woolf once again, there are no beadles at the library door requiring a letter of introduction (1929/1977, 9). Podcasts are communication phenomena that don't require a ticket of entry. They are embodied communication practices that perform on demand from the moment the listener presses play on their digital device.

Embedding podcasts with permaculture content suggests the potential for deep, intimate, and affective, and rebellious acquaintance with the "some place" that Plumwood urges us to make a deep acquaintance with. Tending a garden, particularly the sorts of food producing gardens common in permaculture practices requires the gardener to make a deep acquaintance with place.

> A gardener toils to create beds, unearth roots, plant seeds and turn compost. It can be a bloody occupation: thorns scratch limbs, an ungloved hand risks the stinging weed or legs end up on the wrong side of a sharp shovel or space. This sweat and tears suggests a bodily connection to the land. (Mickelburgh, 2020, 149)

Permaculture is an Australian-developed philosophy which emphasises care for people and the earth, and reduced consumption. It becomes a deeply spatial, temporal, physical and imaginative acquaintance when it merges with the podcast medium. In and of itself, permaculture is not specifically a home garden practice, but the garden (often situated around or near the home) is an essential and integral part of permaculture. Permaculture is a set of design principles focusing on the interconnection of, and cooperation with, nature to create sustainable food and energy systems. These principles, or design systems, facilitate human cooperation with, rather than opposition to, nature. Permaculture places a strong emphasis on small, local practices and individual agency, but also ties these practices to broader social change (Stevens, 2009). There is a sense within some, but not all, permaculture discourse of a re-turn to a historical food producing practices. The practices and stories that emerge from the permaculture soil bring forth a sense of history, gesturing towards a new-old method of communal living. A significant amount of permaculture practices revolve in and around the home, and there is often a rhetorical link with the practices and understandings of urban homesteading, new domesticity, and environmental justice (Padilla Carroll, 2016). The practice and philosophy of permaculture was once considered an unconventional style of living, but both traditional and social media has brought it to a broader, more mainstream audience searching for ways to reject capitalist structures linked with environmental degradation and social injustice. Today many gardeners integrate permaculture practices in their gardens and home spaces.

I chose four *Pip Podcast* episodes, which aired between 2017 and 2018, to consider as I wandered towards a re-turn to environmental communication. The podcasts contained the voices of six Australian women gardeners, being interviewed by Pip editor Robyn Rosenfeldt. When I interviewed Rosenfeldt in 2019 she did not feel her publication was a specifically feminist work and stated she did not exclusively interview women for her podcast. She said:

It's actually not, it's just my world. I create a magazine about what I'm interested in and what people I know are interested in and the people who I'm in contact with. And a lot of them are women. And it's women that are doing things...But there are a lot of women and it focuses on women as much as men, sometimes maybe more than men but I think maybe that's just because I'm a woman. But it's not a conscious thing.

Rosenfeldt's recorded conversations with Australian women living permaculture lives linked permaculture closely to everyday environmentalism through permaculture's three core tenets: "earth care, people care and fair share" (Rosenfeldt, 2017b). I closely examined episodes entitled: "Women as Changemakers" (2017a), "What Is Permaculture?" (Rosenfeldt, 2017b), "Mariam Issa" (Rosenfeldt, 2018a) and "Jodi Vennitti" (Rosenfeldt, 2018b). The words of care and the sense of compassion embedded in permaculture rhetoric and practices are important because they are words and ways which link the "individual actor with broad social change" (Stevens, 2009, 81). Women's compassion for environmental suffering is often entangled with compassion for their human community.

Care and compassion are the action and emotion that matter in gardens and these themes were clearly evident in the podcasts I listened to and studied. Care and compassion, says Mary Phillips, "enable us to visualise the suffering cause by injustice and to consider how best to ameliorate that" (2016, 477). Care is the emotion of compassion, in operation. The Merriam-Webster dictionary defines compassion as the "sympathetic consciousness of others' distress with a desire to alleviate it" (Compassion, 2021). One *feels* compassion, one *does* care. Plumwood specifically pointed to women's stories of care when she spoke of the potential in "new, less destructive guiding stories" to remake the master narrative (1993, 196). The stories I listened to, under Plumwood's guiding words, suggested a communication in common that bloomed through stories about food and flower gardens and lifestyles that embraced waste-free living and reduced over-consumption. These stories suggested a potential to reweave oppressive webs and risk critiques of a return to domesticity and all that free labour and oppression that comes with it.

Care and compassion *are* complex. There are feminist consequences when women talk about, and do, care and compassion work. These words required me to ask difficult questions about the demands made of women through their caring work in, and emerging from, their gardening lives. For Berlant compassion is "a social relation between spectators and sufferers, with the emphasis on the spectator's experience of feeling compassion and its subsequent relation to material practice" (2004, 1). Compassion as "emotion in operation" (4); it is also a word filled with privilege. One must have the resources to alleviate someone or something or someone's suffering. The suffering is never here, it is always "over there" (4). Considering the emotion of compassion and the practice of care from a nourished ecofeminist perspective raises questions about the gendered distribution of power, between who is doing the suffering and who is doing the caring.

These words—*the C words*—have a history of cornering women, of containing them, of re-turning them to the home corner to domesticate and tame. Feminist scrutiny of compassion and care, pays attention to the "cruel optimism" (Berlant, 2011) embedded in these words. Elizabeth Spelman warns that compassion, "like other forms of caring...may also reinforce the very patterns of economic and political subordination responsible for such suffering" (1997, 7). When compassion becomes what Berlant terms a "collective [norm] of obligation" (1), then women embracing care ethics—such as those underpinning permaculture, gardening, and ecofeminist philosophies—also live with a high workload in order to alleviate the suffering they see and feel on both a local, but also global, scale. Sherilyn MacGregor argues that positioning feminist environmentalism amidst an ethic and practice of caring does little for the emancipation of women and the environment (2004, 2006). Failing to recognise the gendered politics of caring that underpins feminist environmental activism "is dangerous if it affirms rather than challenges the feminisation and privatisation of caring work" (MacGregor, 2006, 67). How, she asks, can environmental crises be solved through continued reliance on women providing significant unpaid labour? How, she asks "is this feminist" (2004, 57)? These are important and ongoing questions made of the garden and the women who tell its stories. This reliance and embrace of women's care risks failint to challenge capitalism, the very structure that it seeks to resist, and re-affirms traditional, and often oppressive, gender roles.

Lauren Berlant writes, "there is nothing clear about compassion" (2004, 1) and here too, with the Pip Podcasts things can, at times, be unclear. Val Plumwood admits that the problem with care and compassion, is that it has been "increasingly inexpressible in the public 'rational context', a context that is defined against the domestic sphere in which care has been confined (2002, 35). Yet Plumwood also write that when relationships of care and friendship to other places and the more than human become part of one's identity, these relationships become places of action. I re-turned to the path Plumwood carved almost thirty years ago in Feminism and the mastery of nature (1993) when she said, "where interests are essentially connected and you desire someone else's flourishing for their sake, what is involved is not abandoning your own interest, because in pursuing the other's interest you also pursue, *non-accidentally*, your own" (1993, 151). This relationality not only benefits the carer, but also the cared for—they are no longer treated "as interchangeable commodities which can be chosen or abandoned at will" (155). I realized care and compassion needed re-consideration (perhaps re-calibration). I suspected that when they circulated in the garden and connected through podcasts—when they took up *space*—they might allow us to hear other ways to live, ways that might signal a mutual flourishing. I wanted to consider them not as one or the other but by cutting them together apart (Barad, 2014, 168) and wonder about ways communication might breathe new life into them.

Women as Changemakers Communicating Compassion

She should be rushing. She should be trying to do as much as possible before the cold sets in for the evening. Her body mimics the steady, humble growth of her garden. She knows this common land like the back of her hand. Its methodical growth cycle mirrors her story-telling garden work: both are measured and cannot be hurried. She cannot read or write so she communicates her common garden world by telling. Her story is sound, and her sound is voice, but her words are not written down. It will take the historians centuries to figure out what she says.[1]

If they had been able to read the telling of her un-written story, they would have realised that she and the other women she worked with were "custodians of memory" (Whyte, 2011). *Her garden story in-sisters on un-forgetting the path where the household and neighbourhood interests intersect. Her words fused together with other women like the roots of the plants they grew. Some will dismiss it as wild rumour or gossip, but it is hard to avoid the hints that suggest that through talking and walking and working that she negotiated*

---

[1] The italicised sections within this chapter represent an imaginative "creative history" of the English commons. I hold them close in my analysis of the Pip podcasts, to better understand the importance of communication and communal forms of living.

neighbourhood power. It was the maintenance of knowledge, memories, custom and everyday life.

The "Women as changemakers" (2017a) episode tells of a return-to-the-home environmentalism full of care and compassion. The three Australian women permaculturalists—Su Dennett, Kirsten Bradley, and Meg Ulman—tell their story of care, care-fully to host Rosenfeldt. Their care work is more than garden work and it is more than just the physical, reproductive or groundwork situated in and around the home. When their care work becomes sound-story work it becomes affective work that stretches from the home into relationships and community connections. By verbalizing these emotions in actions, is a story of "affective labor" (Padilla Carroll, 2016, 60). Rather than representing women as returning to a subordinate, domestic role, these women storytell their home gardens and domesticity as political acts that reject the destructive forces of capitalism and the engage both the self, and the community in an "economics of reciprocity rather than profit" (60). Their words of compassion and care ethic are entangled in their good life. It is a good life which is not necessarily an easy, care-free, life. Perhaps, rather, it is a care-full life. That is, it is full of care. These stories sound full of care, but the women speaking them also take care in the way their care is carried out, distributed, and shared.

At the time of writing this book Dennett lived on Melliodora, a property in Hepburn, Western Victoria, with her partner, permaculture co-creator David Holmgren, and son. Dennett and Holgrem run guided tours of their property, and permaculture design courses. They also host volunteers throughout the year. Bradley is from Milkwood Permaculture, a permaculture design business she runs with her partner Nick. When she appeared on the podcast in 2017 Milkwood operated out of Melliodora, but at the time of writing this book the family and their business had relocated to Tasmania. Milkwood offers permaculture design courses, and other related services. Ulman is part of the permaculture blog Artist as Family and lives on a quarter-acre plot in Daylesford, Victoria. The three women told the podcast that they had multiple caring responsibilities: food producing gardens to ensure regular harvest; family and volunteers staying with them; community garden and related projects; domestic animals; and home schooling and other business interests.

As they speak "Women as changemakers" sound out the complexities of living a care-full life in and around their home gardens. Their deep acquaintance with this garden place is the space where a commitment to sustainability and ecological attention is entangled with a full caring work-load. Their bodies are constantly busy caring for the human, and more than human world. On the podcast there is no hiding, or green-washing this fact. It is the first issue Rosenfeldt raises in the podcast, and one that centres much of the almost hour-long podcast. Rosenfeldt re-turns to the issue of caring time regularly throughout the conversation. Ulman explains in the podcast that her family does not use cars or planes, shop in super-markets or purchase general everyday consumer items. They buy second-hand where possible. Significant hours are spent in their garden or preparing home grown food. Both Ulman and Dennet reveal that they have, or had, home-schooled their children. Time, Ulman admits, is an issue they always consider care-fully: "When you are homesteading, when you are making a living from scratch, everything takes time, and every-thing is a lot of work" (Rosenfeldt, 2017a, 16:30) Dennett too describes herself as a "homesteader", defining the concept as a "multi-tasker par excellence" (Rosenfeldt, 2017a, 5:05).

The homesteading the trio refer to is a reference to focusing on the home to reduce reliance on destructive, capitalist systems and focusing more inward, on the home space, for a livelihood (Hayes, 2010; Padilla Carroll, 2016). There is a romantic risk to this homesteading rhetoric. Romance evokes feelings, but it also has the potential to hide realities. Words soaked in a romantic, nostalgic imagery flow from bodies into air-ways as the "Women as changemakers" speak. The romance of home-steading is a rejection of modern consumerism. That is, the purchasing of goods and services that have a direct impact on pollution, land clearing and, often, poor working conditions and low incomes for those manufac-turing the goods. Instead, romantic nostalgia inherent in urban home-steading emphasises a self-sufficiency that re-turns the listener to a simpler life. But as any-one who has ever been in love can testify, romance is not linear. A romantic ethic has been criticised for centring the self and indi-vidualism rather than the community. It emphasises one's uniqueness of peculiarity rather than the features shared with the rest of the human world (Campbell, 2018, 270).

Romanticism also risks glossing over the dual issues that oppress women and other oppressed human groups, and the more than human. Critics argue that romantic ecologists often ignore these twin oppressions and, in some cases, reinforce them. They argue that there is often veiled animosity towards women "under a silk cloak of idealism, protection, and a promise of self-constraint" (Chaia Heller, 1993, 220). In essence, critics say romantic views of homesteading risk glossing over the real work required for such endeavours. The lifestyle associations of gardening, as a place to spend disposable income similar to gourmet cooking and travel, often ignores the physical reality of creating and maintaining a garden (Cadieux, 2013, 62). To garden requires practical, hard, messy physical labour. An emphasis on the romance of self-sufficiency risks sidelining the productive, necessary but incredibly time-consuming caring work that occurs in and around the home and obscures the patriarchal structures that oppress women. It also runs the risk of privileging the white middle class; often those with access to land and the ability to manage and maintain it (64).

Through their conversation the "Women as changemakers" (2017a) acknowledge their high workload, but insist on their agency around their situation. Dennett, for example, does not see her work as a sacrifice, insisting instead on the importance of "looking after yourself first and foremost then your family and then on top of that the community." Unless you do that, she says, "you're not going to be healthy…I see that as I'm not going to be healthy unless I do all three of those things". The "Women as changemakers" podcast (Rosenfeldt, 2017a) emphasise self-management, but seem to resist sliding into a romantic nostalgia for self-reliance that scholars like MacGregor problematise. They highlight their reluctance to allow strangers or even close friends access to their time free of charge. Their time is as valuable as any cash in the monetary economy. They do not reject requests for their time outright; rather, they encourage those requesting it to provide some sort of service in exchange. Ulman discusses the importance of mixing family priorities with creativity and community. She admits,

> Whether its community work or paid work there is sacrifice. So I think its what you're happy to sacrifice and feel ok about in yourself. So what are you going to give up but be ok with in the rest of your life. (Rosenfeldt, 2017a, 18:13)

The "Women as changemakers" sound out garden-home-work-stories that require a significant unpaid domestic workload. Some might perceive this work as a burden; it could be described as suffering, that it reinforces the hard and hostile binary "that defines feminism and homemaking" (Padilla Carroll, 2016, 51). While the sufferer, the environment, is over there, the sufferer, to a listener, might also appear to be the women themselves. Yet the women don't make sounds of suffering. Their words are full of care, but they are also care-full with their words.

Within this context care is not framed as suffering. Dennett explains that she prioritises her home-work time while meeting the time requirements of her guests. Rather than just sitting around talking, guests are invited to assist with the numerous chores and activities that need doing on the property. She says, "that will sort out the sheep from the goats very quickly because they either kind of go yes I'd love to or they kind of go oh well it's time I went". The speakers actively frame their caring load as one they can control and modify for their own, self-care. Dennett, for example, tells a story that extends a care ethic first to herself, and then to building bonds in her family and community. Here, compassion is caring for the self as *part of* caring for community. This is a new way telling of the compassion story about who suffers and who might risk suffering. These narratives articulate ideals that work against patriarchal dominance. Such stories reject the patriarchal, capitalist push for never-ending productivity and economic expansion. Dennett explains:

> You also have to learn to have stronger bonds in your community with whom you should rely more instead of less. In that respect you're substituting your community stuff for the money aspect of security. You're getting a community security instead of a monetary one which does not guarantee anything in a community. (Rosenfeldt, 2017a, 19:16)

While never self-identifying as ecofeminists, Ulman, Dennett and Bradley's words align with ecofeminist philosophy that insists women's home-time is valuable (e.g. Diamond & Orenstein, 1990; Mies & Bennholdt-Thomsen, 1999; Mies & Shiva, 2014). Their caring work is full of worth; their caring words suggest that compassion must first be extended to themselves. Their story-telling of care and compassion does not incorporate the capitalist "glorification of paid work", which has sidelined the importance of non-paid work in the home involving care and compassion (Stephens, 2012, 17). Their non-paid work—the work of care and compassion—is front and centre. It is the only type of work. Their work might often be unpaid, yet the three women have developed practices to ensure it is valued. Bradley, for example, thanks Ulman for showing her how to utilise a bartering system to address problems of overwork and requests for her labour. This can involve planting trees and weeding and cleaning out the chook shed, in exchange for learning new skills. They describe their work-for-time bartering system as "powerful", "beautiful" and "valuable" (Rosenfeldt, 2017a, 32:20). They story their home and gardens as spaces of power and work, and not domestic oppression and drudgery. Bartering is a bodily practice, a cashless economy. Payment occurs through labours of care towards them rather than asking it of them. The bartering system represents the high value they place on their caring work, and caring time. I am intrigued by the way their lives are full of care, but they don't speak words of domestic drudgery, nor of oppression. Instead, this care work is often represented as empowering. Ulman states her family's motivation for choosing their way of life: "it's freeing work, it's mindful work, it's joyous work, it's a celebration of living, it's not being a wage slave kind of work" (Rosenfeldt, 2017a, 16:38). This way of life, she tells listeners, has allowed her family to reduce their reliance on the monetary economy by about 60 per cent. Their workload is high, yet they do not submit to every request asked of them. These additional ideas about an ethics of care have implications for understandings of compassion.

Joyfulness, secreted away amongst care and compassion, emerges willfully in the stories I study. The podcast is a line connecting me to them and I find myself lingering on-the-line, noting that there are other words, words that feel, that cling to care and compassion and give sense to this collective garden story. I sense reasoned argument is not the only thing that sustains their home-working garden lives as their laughter flows down on-the-line. They speak words of care and compassion, but these emotions are nourished by positive, pleasurable alternatives (Johnston, 2008, 103).

Laughter is a medicine, but it is also a fertiliser. There is a comforting closeness in the way these words are surrounded by the sense of Ahmed's affective work, and the need for wonder within feminism. Feminism, Ahmed says, cannot be reduced to being against something, it must also be for something. When something out of the ordinary happens within the ordinary, it creates wonder and bodies that move (2004b, 180). A willfully joy-full female friendship flows from the speaker to me, the listener, rendering compassion practices visible and strong. There is no hero in this story, but there are everyday people with words that are usefully being shared past the borders of their private lives. Dennett explains what this friendship is for, the importance of its use. She says:

Dennett   I think working with women, I love working with men as well, but it's very special to work with women especially who feel that they are in themselves and they are themselves a

	change of something. That we're all living something that we breathe and smell and feel all day everyday.
Rosenfeldt	Yep
Dennett	Even though sometimes it gets you down (laughs) when (inaudible) I just really appreciate the women in my life.
Ulman	I appreciate you.
Dennett	I wouldn't be without them (laughs) (Rosenfeldt, 2017a, 41:00).

The word feminism never raises its head in this pod, but the sound of female friendship is strong. The words of friendship the "Women as changemakers" (2017a) use breathe life into their stories and send stories of a caring connection into the digital world. They are stories that could also nourish ecofeminist theory. Their words within their caring work are stirring a messy mix of love and compassion for each other, but also others in their community. As the women speak, they seem to resist an everyday environmentalism that exploits their bodies. Mackinlay writes of thinking of feminist friends who chase after her with care and compassion (2019, 8) and here to there is a sense of friends caring for each other on their permaculture journeys. Friendship can also be method, when relationships become the cornerstone of research (Mackinlay & Bartleet, 2012). For Mackinlay the word friendship suggests "social affinity, intimacy and trust between women" (75). In the "Women as changemakers" podcast the speakers re-turn to friendship, using it both as environmental method, but also as communication method. Mackinlay unearthed Lillian B Rubin's words about friendship; that it evokes "the best parts of ourselves" (1985, 40) and in the "Women as changemakers" podcast female friendship is evokes and responds to the best parts of compassion in action. They tell a gardening narrative which embraces a female friendship, one which insists on a caring, compassionate relationship.

The "Women as Changemakers" (2017a) story-tellers are clear about compassion, and care as the "emotion in operation" (Berlant, 2004). The storytelling nature of the podcast connects speaker with listener, disrupting understandings of who, and where, the sufferer is. The women do not tell a story of their suffering, but rather one of rejecting structures that cause suffering. This soundwork of care and compassion dissolve from the private friendship of the women speak, into the public sphere via the digital medium of the podcast. If, as Ahmed says, "loving connections are live connections" (2017, 82) then these loving, live connections not only connect the three women but, via the digital technology of the podcast, connect them with a greater collective. As they apply metaphors of love to their spoken tales, these emotions circulate into broader circulation. As the women speak their willfulness passes through and between them like an electric current, switching them on, lighting them up, bringing them together. They are not going with the flow, but rather recognise a willful stance within each other (2017, 82). They are being willful in ordinary places, but their emotions are being carried from their home and their individual bodies and into a collective, listening audience. Ahmed says emotions "bind people together" (2004a, 119) and here too a complex entanglement of care, compassion and willfulness binds the women together, but also weaves itself into the digital world. The energy from this spark brings the women speaking into relation with each other, but via technology it also brings them into connection with the listener. As the listener, I feel similarly charged.

The "Women as changemakers" (2017a) podcast sounds out a commitment to environmental justice. This commitment means a high, often unpaid caring workload, but the storytellers do not define their work as a life of drudgery and submission. This nuanced approach is an alignment with a type of "new domesticity", rejecting both the "masculinized public sphere for wages, but also "[remaining] in the home as a subordinate" (Padilla Carroll, 2016, 56). For the "Women as Changemakers" compassionate practice is not an oppressive obligation. Considering this care work care-fully also allows them to embrace of the joyousness of everyday environmentalism. It is a caring story-line with emotional travelling companions: pleasure, happiness and joy. Maybe not all the time, and perhaps not in all their work, but pleasure appears to often sustain them. Their words are full of will, as well as full of care. They are the willful women who are unwilling to "go with the flow" (Ahmed, 2017, 82). They do not just acquiesce to other's assumptions that their care is an ever-flowing river. And they do not take this stand in anger, but rather in joyful solidarity. They work compassionately within their ordinary spaces but, together, their ordinary gardening lives are the connection with impact that Kathleen Stewart speaks of (2007, 128). Their bodies, united, are affecting one another (Stewart, 2007, 128). But their bodies are also voices that are connecting with others through the podcast. The medium connects their emotions to those listening in.

There's No Place Like Home

Compassion in these words becomes a story about alleviating the suffering of both the self and the community. Perhaps also, it is about alleviating the suffering of the non-human as well. In another Pip podcast "What is permaculture?" (2017b) Hannah Moloney expands on understanding the entanglement of compassion and care ethics with pleasure, joy and love. For Moloney, the subject of this podcast, compassion in the home garden is also compassion in community. She says,

> So, a good life to us is one where we have a vibrant happening home, and that's inside and outside of the house and a vibrant and happening community which is outside our fence line but within walking distance and bike riding distance and that's based on having good relationships mostly so having good relationships with our neighbours, our friends, our community. That's so important to us, it's something we're always working on. (Rosenfeldt, 2017b, 16:01)

Moloney's garden story emerges from her home garden, but revolves around care and compassion within the wider, local environment and community. For her permaculture requires a commitment to analysing how all the systems in her life and work to create "a holistic approach to living on earth". Passive water management and access to her garden, which is situated on a steep, sloping hill, are integral to this pleasure. Rather than toiling up and down a hill, she and her small family have designed paths that contour around their food-producing garden. The garden has been designed to catch water and ensure that it seeps slowly into the soil's top layers. Moloney suggests that through such design much of the garden does not need watering: "That's the kind of thing you can get right when you're designing and leads to many, many, many happy years of pleasurable yard work as opposed to slogging it out…We love our hill" (Rosenfeldt, 2017b, 12:04). Moloney's "kind of thing" seems re-turn to the care-full gardening practice outlined by the "Women as changemakers". She and her family have designed their garden in a way that is kind to themselves. This self-kindness allows garden work to be pleasurable and full of joy rather than laborious and detrimental to their bodies.

The homesteading narrative re-emerges in Moloney's storytelling soundwork. This is her glocal story. She tells Rosenfeldt that her focus is on "relocalising" and taking a "bioregional approach as opposed to a global approach". Her words about the ordinary, everyday, local space, travel from her home, through the medium of the podcast, and into the global space. This too is in sync with the concept of the "radical homemaker", or homesteading, narrative. Drawing on the *Radical Homemakers* work of Shannon Hayes (2010), Moloney understands that her home garden is both a private place full of care, but also a site of empowerment and social and environmental change. Her caring does not focus solely on private problems, through private gardens. It is a sense of compassion for local places and people who are themselves feeling, locally, the repercussions of global environmental degradation. These issues could include resource management, landfill, energy pollution and food and animal protection and safety. For Moloney, pleasure in the garden is also about pleasure in everyday environmentalism. Gardening is more than the physical act; it must be a positive, pleasurable experience. She communicates the story of a good home life, and a good relationship with her community. Listening in there the sense of community dissolves from her private space, into her local community but then into the wider listening world. Le Guin's words about listening being a connection rather than a reaction seem to echo on refrain through the sound of Moloney's voice.

As I listen, I get the sense that community is not just the physical community around her. She sends her story along the airwaves through the podcast, as well as her writing through her website, suggesting her communication as community stretches far beyond her Tasmanian home. Her compassionate feeling turned caring home-work shapes both her home and her connection to others. This connection is pleasurable rather than simply an additional workload. These words, down the line, expand her local to become global.

Compassion lingers in and around these conversations that wander around the home, in the home but also, via the podcasts, out of the home and into other homes. Moloney's home-work is groundwork that becomes soundwork. She is working from home and the home is, is where Ahmed insists feminism lies (2017, 7). Moloney speaks of home as a word that refuses to sit still. It jumps around, chattering in unusual places and difficult places, mingling, and socialising inside and out. Moloney tells of her family's move from Melbourne to Tasmania "because a good life to us is one based in the home" (Rosenfeldt, 2017b, 15:01). She sounds out the word home as her story continues. They could afford to purchase property in Tasmania for this "home-based life"; their work emerges from their home, it is, "a working home, which we love" (Rosenfeldt, 2017b, 15:58).

Home though is based on, or perhaps defined as, relationships: relationships with family and community. Presenting her garden home-work as soundwork Maloney invites listeners to share in the intimate moments of her home-work. She is building relations through this soundwork about home-work. Her story re-turns me to Ahmed's words about feminist homework, of bringing feminist theory home "because feminist theory has been too quickly understood as something that we do when we are away from home" (2017, 8). But Moloney's words also suggest the desire to generate intimacy with her listening public. Moloney is setting a scene. She is narrating an experience of belonging and intimacy that revolves around her home garden. This scene mingles from the private to the public, circulating certain expressions and desires (Dobson et al., 2018, 5). For the listener the sense of home expands because it has travelled from Moloney's home to those who listen to her stories. The listener remains a stranger, but the medium provides a welcome into her home.

Moloney tells the podcast that she is informed by the notion of radical homemaking outlined by American writer Shannon Hayes (2010). Hayes examines the complexities involved when individuals and families wish to care for the environment through reduced consumption, increased connection to home and community, and eating and growing food locally. Hayes too grapples with the implications of a return-to-the-home, acknowledging the importance of the gains made by the feminist movement in previous decades and the risks of a romanticised nostalgia for a previous era. Compassion, in the form of Haye's conception of radical homemaking, is risky writing; it risks the continued reliance on free and feminized labour of which Sherilyn MacGregor warns. Yet Moloney seems to resist this resistance narrative. Instead her communication appears to follow the way forward outlined by sociologist Laural Richardson and her insistence that communication should be about the "communion we can create, not the hegemonies we can resist" (1997, 79). Moloney insists her radical homemaking involves "bringing the heart back into the house". Her words of care and compassion emerge from the home and centre around the home. For Moloney, home is a private residence, but it is also community. Her garden story is more than a place of harvest and it is more than a place of home. It is part of a larger system of living in the local and global community. Her caring work happens in the place she calls home, but her compassionate breaches those spaces, crossing airwaves and oceans to connect with other hearts and minds.

The podcast, "Jodie Vennitti" (Rosenfeldt, 2018b), follows a conversation between Australian single mother Vennitti and Rosenfeldt. Her voice on-the-line details her one-woman, year-long experiment in eating only what is produced in her garden. Caring again is front and centre in this garden work, but it is one that speaks the language of mutual flourishing. She reveals multiple responsibilities of being single mother, studying for a degree and struggling to find time to harvest and eat her home-grown produce. When she quickly realised she couldn't do it all on her own, her garden work breached the privacy of her home and searched for connection. Vennitti connected with her permaculture group, the Freo Permies, for support. The Freo Permies often help each other out in a fair and equitable way, often via working bees, but not through a formal or obligatory system. Rather, suggests Vennitti, the notion of working together, helping each other out, or donating goods is done for emotional reasons of pleasure or joy and a sense of community and connectedness. There is a sense of mutuality in her relationship with the Freo Permies; it clings to her compassion story and caring work. Vennitti says that the Freo Permies know her garden as intimately as she does, and she shares stories as well as garden produce with them. This mutuality is a sense of a re-turning to bell hooks words about the generous sharing of resources being a "concrete way to express love" (2001, 163). It is through friendship, says hooks, that both community and love are built. It is here where care, respect, knowledge, and nurturing occurs (2001, 133).

The garden here becomes a space of public-private entanglement. Working and growing her garden brings her words and life into communion with her neighbourhood, but also to listeners. Her soundwork includes storytelling her free food project where she places a cart, loaded with free fresh food grown from her garden, on her verge. The cart and its position on the verge in this story becomes a hybrid space. Legally, the verge is a public space, but her bodily action and words turn this into a space shaped by her actions and the responses of her neighbours. Vennetti greets strangers as friends via a Facebook group to alert people about her free verge food. She also grows some food on her verge, and though council laws stipulate restrictions to her verge plantings, she decides that the benefits outweigh the risks and proceeds to garden there anyway.

There is a collectivity to this individual story as she and her neighbours determine their own ways to best use a public space, that sits adjacent to her private land. Vennetti emphasizes that the food she grows both in her home garden and on the verge, as well as her home-grown food place in the vegetable cart "is for everyone". Anyone, no matter where they live locally, regionally, or globally, can read about and view (both literally and digitally) the food and flowers on her front lawn. There is no exclusion here. Vennitti speaks about the garden being connection with her neighbours, and others from further afield. She says:

> There's one lady who comes every Sunday morning at 7:30 and she now comes in and we have a chat and a yarn and she's become a really large part of my life, and mine hers and she takes a lot of the produce and she turns it into preserves and pickles and it comes back in a jar. (Rosenfeldt, 2018b, 30:32)

Through her groundwork as soundwork Vennitti tells stories about connections with friends and strangers. bell hooks' says a requirement for a living, loving community is meeting strangers without fear and extending to them the gift of openness and recognition (2001, 143). Vennitti storytells the expansion of not only her garden borders, but also what constitutes her community. There is no sense of a "collective norm of obligation", that Berlant worries compassion entails (2004, 1). It is the communication of community through communion. Giving, for Vennitti is front and centre in her communication and giving, as hooks says, "brings us into communion with everyone" (2001, 163). Vennetti's free garden food is a compassionate act of both empowerment and communion. It is also compassion which requires rejection. She rejects the enclosure of her garden. She rejects a fear of strangers. While she doesn't overtly reject laws that restrict access to her land, she places connections with community as a main priority. She says:

> I think that's an important conversation because we have a society that we don't have that connection, we have high anxiety, depression, people feeling really lonely and busy in cities and if we can do things to help each other and have something beautiful just around the corner from you I think we certainly need that as a society. (Rosenfeldt, 2018b, 31:29)

Merging her groundwork with the podcast sound work, Vennitti stretches the boundaries of her garden space from the privacy of her home to the broader community. As she takes up sonic space through the podcast, the physical boundaries of her garden space become fainter, almost invisible. Her physical garden is still situated around her home, but the storytelling generates a sense of "public intimacy" (Dobson et al., 2018), where words of care and compassion which emerge from private ground, creep across the boundaries into the public world.

FOLDING DOWN THE SILENT SWAG

Even in soundwork, silence can remain. It is important to remember that absence also has presence. Radio silence is not dead air; it is also part of a story. There is privilege in having a garden to tend in the first place. There is privilege to viewing suffering—ecological change—"over there", from the comfort and safety of a home that provides both shelter and sustenance. The white middle-class voices I listened to have access to land, and in Australia land sits in a political, historically violent soil. And so, the question must be asked; who is absent from these stories? Who does not have the privilege to tell a compassionate garden story? Rosenfeldt does approach the issue of privilege in one of her podcasts when she asks the "Women as changemakers" (2017a) about the relationship between their ability to be self-reliant and their privilege. Bradley responds, acknowledging her privilege and affirming that she does not have to face discrimination based on race, nor is she a single mother, nor does she have many children. She says:

> I'm not starting from a starting point way back behind where, as a white educated female with parents that were able to stay alive and love me. It's incredible the amount of different points that we start from and I think when we're busy in our lovely lives and we're going, oh it's so hard I have to preserve all these delicious cherries I just picked. I think it's often very, very important to get a grip on how bloody lucky you are to be in this position at this time on planet earth. (Rosenfeldt, 2017a, 44.24)

Ulman too acknowledges her privilege and ways she gives back through practical action. She admits to having "great guilt that goes along with that privilege". Her method of allaying some of that guilt is to work with community to bring more garden structures to the area. She says:

I think starting up the five community gardens we have in Daylesford is a
way to lessen that guilt and to share some of that privilege of having free
organic food. I think that's part of it. Yes, to have an exchange but also to
offer a lot of workshops for free that includes everybody who may not have
the five bucks or whatever it is to come along. (Rosenfeldt, 2017a, 45.13)

There are clear attempts, within these conversations, to consider privilege.
The women who speak acknowledge their privileged compassion story. Yet,
while the true sufferer remains "over there", as Berlant would say (2004, 4),
suffering has also occurred under there and around here, under the soil that
is gardened on and the sounds travelling across the airways. The silence
about the violent history embedded in the Australian soil where gardens are
grown remains subsumed in this compassion story of home-making and
home-steading. Aileen Moreton-Robinson's words, that for Indigenous
women "all white feminists benefit from colonisation" (2000, xxv) rings as
true now in this soundwork, as they have in the long history of Australian
women's garden stories. The "white possessive logics" Moreton-Robinson
speaks about Australians benefitting from, remains here too, in these podcasts (2015, xii). Belonging in these home places, no matter what compassion work circulates, remains based on homesteading dispossession of the
original owners of the land. The soundwork of the women I listen to tell
important stories about care and compassion, about community and coming together. Yet silence remains within this compassion, silence about the
fact the land they work on has been stolen, and silence about the violence
that has allowed others to live and own on this land. Such stories risk failing
to heed Tuck and Wang's warning about urban homesteading constituting
problematic attempts to reconcile "settler guilt and complicity" (2012, 3).
For Tuck and Yang urban homesteading constitutes one of many "settler
moves to innocence", that is "strategies...to relieve settler feelings of guilt
or responsibility without giving up land or power or privilege" (2012, 28).
There is no suggestion within the Pip Podcasts of evoking the kind of
"'Indigeneity' as tradition" that Tuck and Yang point to within much urban
homesteading discourse. Yet the silence around Indigenous sovereignty and
modern presence remains.

While there is hope and optimism within this caring and compassionate soundwork, allowing the silence to sit, un-earthed and un-consid-
ered, fails to do the "hard, unsettling work of decolonization" (Tuck &
Yang, 2012, 4). It continues the white, middle-class monologue. Ethical
dialogue, says Bird Rose, "requires that we acknowledge and

understand our particular and harshly situated presence" (2004, 22). This soundwork as groundwork I listen to occurs on stolen land, on damaged places where settler-colonial violence has happened and keeps happening. Silence about this fact is, as Bird Rose says, "the big swag" that lies between us (2004, 22). I re-turn to the words of Ambelin Kwaymullina who points to the danger hidden in decolonizing dialogues between ecofeminists and Indigenous women (2018). There is danger, she says, when ecofeminists reproduce feminist failures to address intersectional oppression of Indigenous women, and settler women's complicity in this oppression. There is danger in the environmental movement's failure to recognise Indigenous management of cultured Countries. She concludes, "ecofeminists who do not interrogate position and complicity will promulgate rather than challenge the structures that continue to oppress Indigenous homelands and Indigenous women" (197).

It is important, then, to remember the legacy of unpaid care work that risks reproducing conditions of oppression for other women. Ahmed warns that we must observe who is working for whom. Employing other women to free up our time and energy means "we are simply passing our exhaustion on to others" (2017, 86). If being freed from labour requires other women to labour, "others are paying the price of your freedom. That is not freedom" (p. 86). Failing to address the labour required and intensified through increased self-reliance and self-sufficiency blinds us to the impact on the women throughout the world who do more unpaid domestic labour than men, and other women. Not doing so might risk re-turning to "the possibility for power and privilege to be mis/used/represented to perpetuate and reproduce dominance of self in relation to other" (Mackinlay, 2019, 183). Compassion shapeshifts its way around the garden in an ethics of care, and so I keep those historical legacies close. Critical work is legacy work.

Reimagining Women's Compassion Work

Walking these garden paths with friends brings her community into communion. She watches her garden grow and cultivates "care, knowledge, respect and responsibility" both for herself, her garden and all of them (hooks, 2003, 242). *They, the men with the money, call this land the wastes but nothing she grows either in the land, or in her friendships, goes to waste. When these men enclose her common world with capitalist words they forget to talk about her body and waste her words. When enclosure happened words of compassion went to waste. Yet even waste is essential to nourishing garden soil.*

Mariam Issa, a mother and community gardener who came to Australia as a refugee from Somalia, tells a compassionate community story about gardens to Rosenfelt (2018a). Issa tells a story of motherhood made difficult through her experiences of being a refugee, and racial discrimination. Before creating her community garden, it was she who suffered. In her compassion story suffering is not only, as Berlant says, "over there"; it has been "over here" as well (2004, 4). Through her story Issa reveals both witnessing, and experiencing, huge trauma on her journey to Australia. Issa's community garden story begins with a war story. It also begins with being a mother to four, with a fifth on the way, seeking asylum and subsequently settling in Melbourne, Victoria. Her words speak of dislocation; of fleeing from violence and seeking asylum in Australia. Entangled in this story are words of grief, anger, and trauma. Before she even speaks about her garden, she speaks stories of mothers' grief and loss. She says:

> I saw in a refugee tent a woman sharing a story that her child was hanging on to her when her boat capsized and because she was saving another child, she had to literally rip the hands of her child off her shoulder to let him drown. Imagine those women living among us, how do they forgive themselves? So, women are carrying those atrocities, they are carrying those memories within them. And that is where the war will continue and unless women heal then I don't believe that war outside will stop. (Rosenfeldt, 2018a, 3:37)

I take time to consider what I want to say about the affective meaning within her communication. I shed tears listening to her story and hold Barad's words in my palm to re-turn to and reflect on my affective response. If I want to write about connection and caring and compassion I can only start with motherhood because I don't know where else to start. My feminist sensibilities cringe as I equal parts exclude and preclude with those three little words: "as a mother". To say "as a mother" gestures towards some sort of equal experience between she and me, between then and now. It also suggests that those who are not "as a mother" cannot connect, or understand or relate in the way a mother day. And yet, when Issa speaks of mothering, I connect with her thinking about my own connection with my children. I have to start with motherhood because that is what drew me to ecofeminism and I don't know where else to start when it comes to wondering about the communication of care and compassion of a garden story started by a woman who has moved mountains, when I have only grown a garden. As a mother I know about love and loss and sick children and strange places and faces. As I mother I know that a mother notices things because she is there all the time, and most of the time and some of the time. But as a mother I don't know about war and boats capsizing and fleeing guns and babies dying and mothers being human cargo and prejudice because of the colour of my skin. As mother I have never experienced time on a leaky boat. Issa describes that time in the podcast. She says:

> During the daylight, the children and the turmoil and the cries of hungry children and elders was just so much to bear. But during the night it was quiet. And I would look up and I have never seen so many stars in my life and that time I realised that there was a source of power within us that was way beyond what we sometimes feel and are connected with. In the nights I would have this hope that all is going to be well. (Rosenfeldt, 2018a, 7:00)

If storytelling is embodied communication, and if I listen to Issa's story in my home, then my homework is to wonder how words of care and compassion connects her to me and her. I consider words as a mother in my mothering home, the home where I do my intellectual and emotional work (Ahmed, 2017, 7). Issa's garden story is a suffering story. But that is not all to the story. Issa's own suffering might be the ground through which her compassion grows, but as she continues to speak, I sense that it is not what sustains it. While many gardens started by people with migrant and refugee experiences are established in highly visible, vacant public lots, Issa's community garden RAW (Resilient Aspiring Women) grew from her own home space. She care-fully communicates compassion and care as trust, rather than transactional. In her garden, compassion collaborates. Her community garden grows in a private space. Her words, merged with her physical garden make it unclear where her private home space ends and the public begins. Perhaps that is the point of her story. She says:

> The community garden was in my own backyard. I always wanted to work with women and in this time of my journey I realised that women needed healing...And so I didn't want something really huge. I just wanted to bring women together from diverse cultural backgrounds just to have tea and talk and hold that space for them. And incredibly the organisation became something more than that. (Rosenfeldt, 2018a, 21:37)

As Issa explains the development of her RAW (Resilient Aspiring Women) community garden, she seems to suggest that the garden space is a place where she and other members of the community alleviate each other's suffering. The RAW garden, Issa says, began because "women needed healing" and she wished to address what she calls a "trust deficit" in the community (Rosenfeldt, 2018a, 21:24). Compassion in the RAW garden extends not only to herself, but also to strangers. She tells the podcast that women come to the garden for a variety of reasons. Some because they live in small spaces and can't have a garden, while others want to learn how to garden. It is not just women who attend the garden. The space is supported by school students who help build and decorate garden areas. It also incorporates cooking classes, and art and craft spaces. Rotarians built a pavilion which students then decorated, and an organisation called Igniting Change built a kitchen.

There is compassion in this garden story, and one the collective might strive to ameliorate, but within that obligation is also pleasure and joy. Perhaps this compassion is a "collective norm of obligation" that Berlant speaks of when she writes about reading a scene of distress "not as a judgment against the distressed but as a claim on the spectator to become an ameliorative actor (2004, 4). And yet, I sense there is something more going on, in this sonic scene. Through a care ethic built on community connections, Issa grows a sense of belonging in her garden. Her words open a traditionally enclosed, private property, as a public site. The suffering is no longer "over there" but within and between Issa and the community. There is no specific clarification within the podcast about how the garden is funded or maintained; the conversation implies that much of the work carried out on the garden is conducted by volunteers. Issa does not specify precisely how much work she has put into the garden herself, but she highlights what may be motivating the volunteers who are involved in the project: the very act of coming together, sharing and caring. Issa's compassion work is extensive; according to her story her caring load seems high. Yet it is complemented and supported by numerous other organisations and individuals. It embraces both suffering and spectators. The women who attend the RAW garden are from her suburb, Brighton, as well as wider Melbourne.

Sharing the workload alleviates workload, but it also increases the load of pleasure and joy. Issa explains that the RAW garden was created "from a place of trust" where "no other transactions are exchanged". Much of the garden harvest is used for communal and celebratory purposes. There is nothing to indicate that volunteers take the food home with them (although this is possible), but Issa tells us that it is used for communal dinners, group cooking classes, oral storytelling, music and art and craft. By opening what is traditionally enclosed, private property, as a public site, Issa has used compassion to create a new community and community connections, and, in doing so, a sense of belonging for all who visit her garden. She says:

> I think what this community garden has done for us, for myself and my family…is it has really taught us family beyond blood relations and the power of community. In our African cultures we say it takes a whole village to raise a child. I think the garden has given me wisdom enough to add to that saying that when we nurture the woman, we nurture the village. (Rosenfeldt, 2018a, 26:54)

Each year she gets better at it. The soil, and her ideas, become more fertile. Her project of enriching the soil produces fruitful stories. She can feel the tomato growth in her words as she searches for the words that reanimate the world (Plumwood, 2009). She doesn't have time wonder about whether she is speaking the world the right way, if she is running too fast with words. She wraps up her words in a brown paper bag and sets off into the field. There is work to be done.

Listening to Compassion and Communication

I scoop up these words of care and compassion. I weigh them, one word in each hand. They seem lighter now, yet contain greater meaning. Care and compassion are not words that are strung, or pegged up, along on this line alone. Pleasure and joy seem to have lodged themselves alongside compassion, as the emotion of feeling with. I make space in my harvest basket as these words become larger and broaden my thinking about what ecofeminism is about.

These stories-on-the-line—these Pip Podcasts—are full of care and compassion but they are also care-full about compassion. They are strangers on the line but over time the strangeness of their voices becomes more familiar. The digital intimacy emerges in the sound of their voice, and the words on their tongue. A line can be a connection with the potential to build trust between friends, neighbours, and strangers. Their self-less, but care-full word giving aligns with a generosity of spirit that brings them "in communion with everyone" (hooks, 2001, 163).

I am still not sure whether simply being on-the-line makes you part of a community. But this communication comes at a price. Burdening women with environmental care can leave them, once again, shouldering the responsibility for compassionate work. Operationalising a compassionate emotion into caring work risks re-turning to the path of least resistance: more unpaid and subsequently undervalued work. As a feminist scholar it is my response-ability to wonder about the consequences to an eco-feminised commitment to care. The words of the women I study provide one path through this problem.

These are stories stuck on the fenceline. On the one hand feeling strongly about the need for care and compassion in words and ways of being and doing everyday environmentalism. On the other hand, knowing, on a personal level, the huge responsibilities required to un-earth and un-forget the legacy of inequality and privilege that continues to grow when settler-colonial women speak and connect with these words. To consider the way these sisters of the soil spoke words of care and compassion in and around their gardens risked a conclusion that heaped more unpaid caring responsibilities on women. It risked writing which compounded the historically unequal relations between men and women. And it risked, once more, ignoring the secrets embedded in the soil.

Composting care and compassion stories is a messy feminist mix. I feared making trouble, repeating history, by making the maternal foundational to an ecofeminist "alternative politics of compassion" (MacGregor, 2004, 60). I feared trouble with feminist friends who have, quite understandably, worked hard to distinguish women's bodies from nature. When the word care is silent about the exploitation of women and the unequal division of unpaid labour between the sexes, then its definition is incomplete. Yet, much like the compassionate feel obliged to ameliorate suffering, as feminist communications researcher I am compelled to attend to my response-abilities, and that includes the tackling the complexities of compassion (Lipton & Mackinlay, 2017; Mackinlay, 2019, 225). Caring work and compassionate feeling is important for ecological and social justice. Many need to care for something to change, to improve. The ethics of care is not the problem; the key issue is who is doing the caring work and the value placed on that work. Ecofeminists have long pointed to the way the word caring is given low-level importance because, by its very nature, it has a limited capacity to improve productivity. As many a mother knows, there is little that can be done to speed up the feeding of a newborn infant or bathing a child (Mies, 1998, 221; Wichterich, 2015, 87). Precisely because that seemingly unproductive caring work is carried out by so many women is the same reason it is deemed unimportant. We can learn many things about the ways we communicate compassion from these women who speak their stories. They are stories that reveal much about how to be care-full about communicating compassion and care.

It's a stretch to say that just listening turns us all into friends. They don't know that I am listening, and we are still strangers to each other. I am re-turned to Palyku writer Ambelin Kwaymullina call for non-Indigenous ecofeminists like me, who continue to benefit from the dispossession of Indigenous women form their homelands, to start listening. Within the Pip podcasts I have chosen to analyse I am only listening to women who are settlers. Yet, sound remains a connection, and it echoes down the line. The absence of Indigenous women's voices should be noted, interrogated, denounced. It is an absence that needs to be filled. Within her own writing Le Guin herself grappled with colonialism and her complicated position within that system. If, as Le Guin suggests, nobody can do anything much alone, then as these voices roll out one after the other along the line, I am listening to those "who have imagined life along lines that make sense and allow some freedom" (2004, 209). My wanderings along this line have led me to wonder about how female friendship communicates compassion in the garden. I re-turn to Lillian B Rubin who says, "people maintain an important emotional link with best friends of the past while also developing new intimate relationships where they live" (1985, 177). If listening is, as Le Guin suggests, an "act of community" then women I listen to are more like a "critical companion" (Mackinlay, 2019, 8). The problem is, there needs to be more companions on the line, ones who can be critical.

REFERENCES

Ahmed, S. (2004a). Affective economies. *Social Text*, 22(2), 117–139. https://doi.org/10.1215/01642472-22-2_79-117

Ahmed, S. (2004b). *The cultural politics of emotion*. Edinburgh University Press.

Ahmed, S. (2010). *The promise of happiness*. Duke University Press.

Ahmed, S. (2017). *Living a feminist life*. Duke University Press.

Ahmed, S. (2019). *What's the use?: On the uses of use*. Duke University Press.

Barad, K. (2014). Diffracting diffraction: Cutting together-apart. *Parallax: Diffracted Worlds—Diffractive Readings: Onto-Epistemologies and the Critical Humanities*, 20(3), 168–187. https://doi.org/10.1080/13534645.2014.927623

Berlant, L. (2004). Introduction: Compassion (and withholding). In L. Berlant (Ed.), *Compassion: The culture and politics of an emotion* (pp. 1–14). Routledge.

Berlant, L. (2011). *Cruel optimism*. Duke University Press.

Bird Rose, D. B. (2004). *Reports from a wild country: Ethics for decolonisation*. University of New South Wales Press.

Bonenfant, Y. (2014). On sound and pleasure: Meditations on the human voice. *Sounding Out*. https://soundstudiesblog.com/2014/06/30/on-sound-and-pleasure-meditations-on-the-human-voice/#:~:text=%20On%20Sound%20and%20Pleasure%3A%20Meditations%20on%20the,Anglo-Sax-on%20standards%2C%20and%20her%20voice%20wa

Bunda, T. (2018). Seeing the Aboriginal sovereign warrior woman. *Lifted Brow*, 40, 4–5.

Cadieux, K. V. (2013). Other women's gardens: Radical homemaking and public performance of the politics of feeding. In A. Hayes-Conroy & J. Hayes-Conroy (Eds.), *Doing nutrition differently: Critical approaches to diet and dietary intervention* (pp. 61–86). Ashgate.

Campbell, C. (2018). *The romantic ethic and the spirit of modern consumerism* (2nd ed.). Palgrave Macmillan.

Chadha, M., Avila, A., & Gil de Zúñiga, H. (2012). Listening in: Building a profile of podcast users and analyzing their political participation. *Journal of Information Technology & Politics*, 9(4), 388–401. https://doi.org/10.1080/19331681.2012.717481

Compassion. (2021). *Merriam-Webster*. https://www.merriam-webster.com/dictionary/compassion

Copeland, S. (2018a). Broadcasting queer feminisms: Lesbian and queer women programming in transnational, local, and community radio. *Journal of Radio and Audio Media*, 25(2), 209–223. https://doi.org/10.1080/19376529.2018.1482899

Copeland, S. (2018b). A feminist materialisation of amplified voice: Queering identity and affect in The Heart. In D. Llinares, N. Fox, & R. Berry (Eds.),

Podcasting: New aural cultures and digital media (pp. 209–225). Springer International Publishing AG.

Diamond, I., & Orenstein, G. F. (1990). *Reweaving the world: The emergence of ecofeminism.* Sierra Club Books.

Dobson, A. S., Carah, N., & Robards, B. (2018). Digital intimate publics and social media: Towards theorising public lives on private platforms. In A. S. Dobson, B. Robards, & N. Carah (Eds.), *Digital intimate publics and social media* (pp. 3–27). Palgrave Macmillan.

Haraway, D. J. (2016). *Staying with the trouble: Making kin in the chthulucene.* Duke University Press.

Hayes, S. (2010). *Radical homemakers: Reclaiming domesticity from a consumer culture.* Left to Write Press.

Haygood, D. M. (2007). A status report on podcast advertising. *Journal of Advertising Research, 47*(4), 518–523. https://doi.org/10.2501/S0021849907070535

Heller, C. (1993). For the love of nature: Ecology and the cult of the romantic. In G. Gaard (Ed.), *Ecofeminism: Women, animals, nature* (pp. 219–242). Temple University Press.

hooks, b. (2001). *All about love.* Harper Collins.

hooks, b. (2003). *Communion: The female search for love.* Perennial.

Johnson, P. A. (2008). The howl that could not be silenced: The rise of queer radio. In M. C. Keith (Ed.), *Radio cultures: The sound medium in American life* (pp. 95–112). Peter Lang.

Johnston, J. (2008). Counterhegemony or bourgeois piggery? food politics and the case of foodshare. In Middendorf, G., & Wright, W (Eds). *The fight over food : producers, consumers, and activists challenge the global food system.* Pennsylvania State University Press.

Kwaymullina, A. (2018). You are on Indigenous land: Ecofeminism, Indigenous peoples and land justice. In L. Stevens, P. Tait, & D. Varney (Eds.), *Feminist ecologies: Changing environments in the anthropocene* (pp. 193–208). Palgrave Macmillan.

Langellier, K. M. (1989). Personal narratives: Perspectives on theory and research. *Text and Performance Quarterly, 9*(4), 243–276. https://doi.org/10.1080/10462938909365938

Le Guin, U. K. (1989). *Dancing at the edge of the world: Thoughts on words, women, places.* Grove Press.

Le Guin, U. K. (2004). *The wave in the mind: Talks and essays on the writer, the reader, and the imagination* (1st ed.). Shambhala.

Lipton, B., & Mackinlay, E. (2017). *We only talk feminist here: Feminist academics, voice and agency in the neoliberal university.* Springer International Publishing.

Llinares, D., Fox, N., & Berry, R. (2018). Introduction: Podcasting and podcasts-parameters of a new aural culture. In D. Llinares, N. Fox, & R. Berry (Eds.),

Podcasting: New aural cultures and digital media (pp. 1–13). Springer International Publishing AG.

MacGregor, S. (2004). From care to citizenship: Calling ecofeminism back to politics. *Ethics & the Environment, 9*(1), 56–84. https://doi.org/10.2979/ETE.2004.9.1.56

MacGregor, S. (2006). *Beyond mothering earth: Ecological citizenship and the politics of care.* UBC Press.

Mackinlay, E. (2019). *Critical writing for embodied approaches: Autoethnography, feminism and decoloniality.* Palgrave Macmillan.

Mackinlay, E., & Bartleet, B.-L. (2012). Exploring the potential of sisterhood and personal relationships as the foundations of musicological and ethnographic fieldwork. *Qualitative Research Journal, 12*(1), 75–87. https://doi.org/10.1108/14439881211222741

Mickelburgh, R. (2020). Compassion in the garden: Radical homemakers or just more women's work? *Emotions: History, Culture, Society, 4*(1), 146–166. https://doi.org/10.1163/2208522X-02010092

Mies, M. (1998). *Patriarchy and accumulation on a world scale: Women in the international division of labour* (New ed.). Zed Books.

Mies, M., & Bennholdt-Thomsen, V. (1999). *The subsistence perspective: Beyond the globalised economy.* Zed Books.

Mies, M., & Shiva, V. (2014). *Ecofeminism.* Zed Books.

Moreton-Robinson, A. (2000). *Talkin' up to the white woman: Aboriginal women and feminism.* University of Queensland Press.

Moreton-Robinson, A. (2015). *The white possessive: Property, power, and indigenous sovereignty.* University of Minnesota Press.

Padilla Carroll, V. (2016). The radical possibilities of new (feminist, environmentalist) domesticity: Housewifery as an altermodernity project. *Interdisciplinary Studies in Literature and Environment, 23*(1), 51–70. https://doi.org/10.1093/isle/isw013

Phillips, M. (2016). Embodied care and planet earth: Ecofeminism, maternalism and postmaternalism. *Australian Feminist Studies, 31*(90), 468–485. https://doi.org/10.1080/08164649.2016.1278153

Plumwood, V. (1993). *Feminism and the mastery of nature.* Routledge.

Plumwood, V. (2002). *Environmental culture: The ecological crisis of reason.* Routledge.

Plumwood, V. (2009). Nature in the active voice. *Australian Humanities Review, 46*, 113–129. https://doi.org/10.22459/AHR.46.2009.10

Richardson, L. (1997). *Fields of play: Constructing an academic life.* Rutgers University Press.

Rosenfeldt, R. (Host) (2017a). #8 Women as changemakers [Audio podcast episode] Pip Podcast. *Pip Magazine.* https://www.pipmagazine.com.au/podcasts/pip-podcast-8-women-change-makers/

Rosenfeldt, R. (Host) (2017b). # 3: What is permaculture? with Hannah Moloney [Audio podcast episode] Pip Podcast. *Pip Magazine.* https://www.pipmagazine.com.au/podcasts/what-is-permaculture-hannah-moloney/

Rosenfeldt, R. (Host) (2018a). #10 Mariam Issa [Audio podcast episode]. Pip Podcast. *Pip Magazine.* https://www.pipmagazine.com.au/podcasts/pip-permaculture-podcast-10-mariam-issa/

Rosenfeldt, R. (Host) (2018b). #11 Jodie Vennitti [Audio podcast episode]. Pip Podcast. *Pip Magazine.* https://pipmagazine.com.au/news/pip-permaculture-podcast-11-jodie-vennitti/

Rubin, L. B. (1985). *Just friends: The role of friendship in our lives.* Harper & Row.

Spelman, E. V. (1997). *Fruits of sorrow: Framing our attention to suffering.* Beacon Press.

Stephens, J. (2012). *Confronting postmaternal thinking: Feminism, memory, and care.* Columbia University Press.

Stevens, S. (2009). The official rhetoric of permaculture: Motivating behaviour change through environmental communication. *Australian Journal of Communication, 36*(2), 73–91.

Stewart, K. (2007). *Ordinary affects.* Duke University Press.

Tiffe, R., & Hoffmann, M. (2017). Taking up sonic space: Feminized vocality and podcasting as resistance. *Feminist Media Studies, 17*(1), 115–118. https://doi.org/10.1080/14680777.2017.1261464

Tuck, E., & Yang, K. W. (2012). Decolonization is not a metaphor. *Decolonization: Indigeneity, Education and Society, 1*(1), 1–40.

Van Djick, J. (2007). *Mediated memories in the digital age.* Stanford University Press.

Whyte, N. (2011). Custodians of memory: Women and custom in Rural England c. 1550–1700. *Cultural and Social History, 8*(2), 153–173. https://doi.org/10.2752/147800411X12949180694263

Whyte, N. (2015). Senses of place, senses of time: Landscape history from a British perspective. *Landscape Research, 40*(8), 1–14. https://doi.org/10.1080/01426397.2015.1074987

Wichterich, C. (2015). Contesting green growth, connecting care, commons and enough. In W. Harcourt & I. L. Nelson (Eds.), *Practising feminist political "ecologies": Moving beyond the "green economy"* (pp. 67–100). Zed Books.

Woolf, V. (1929/1977). *A room of one's own.* Granada.

Woolf, V. (1952). *The leaning tower* (Uniform ed.). Hogarth Press.

PART III

Letter Three (Winter)

My dear K,

The word "verge" can be a noun and a verb: a place and a state of mind. I have been studying pictures that emerge from a neighbourhood verge and on to the screen. It feels like they also verge on another time and place. I write this letter to you backwards in time, in an attempt to catch up with what has happened and what the images I view are trying to make happen. Perhaps this is my attempt to dis-entangled myself from what Deborah Bird Rose calls a "web of time concepts" (2004, 18). Sara Ahmed says, "willing together can be an experience of being in time" (2014, 50) and this case study catapults me through time and in tune with women I have never met, so can only imagine. I am trying to be in time with another time and so I search for some semblance of women's willfulness to re-turn to their common land, at another time. I am unsure about whether I am in time—whether I have got the rhythm right. It feels like I too am on the verge of something.

As I grapple to find meaning within online images of a neighbourhood verge garden Kathleen Stewart's affective words run in circles around my head. She says, "there's a politics to being/feeling connected (or not)...to all the forms of attunement and attachment" (2007, 15). The digital imagery of the project I examine—*Urban Food Street*—seems immediate and close on the screen, and yet in some ways I am still kept at a distance. I am not part of the neighbourhood, and yet the online nature of this imagery attaches me to it. These images prompt me to seek out the post-cards from your Cooloola campaign. Your postcards to save Cooloola were the physical, material matter that could, and still can be, touched, felt, smelled. I now walk, swim and camp at Cooloola, thanks to you. They were tools put to important use. They remain full of use.

Urban Food Street has shown me how garden objects are use-full time travellers. They have a certain use in the garden, but they also have an underlying use. When writing about use Ahmed says "objects can be in time with each other, travelling in time. Objects also travel through time" (2019, 25). Her reminder that use does not necessarily correspond to an intended function resonates clearly when I search for deeper meaning in pictures of shovels and spades, pitchforks, and wheelbarrows. They are ordinary objects, carrying out both an intended and, perhaps, unintended function.

This is the chapter that, more than the others, grew from the echoes of ecofeminists in-sistering that "there is no commons without a community" (Mies, 2014b, 106). I have searched these images for signs of a contemporary reclaiming of the commons. I continue to write and dream my way into a past herstory of English commons enclosure protest, to try to link the thin threads in this web of stories, and time concepts. In *Urban Food Street* I see gestures towards this enclosure protest—a protest women's bodies were entangled in. It is a story about use and form, about the way the suburban form could be put to better use (Ahmed, 2019).

Again, much like the other stories I study, an absence remains within these images I study, and I shift uncomfortably in my chair as I grapple with this unsettling knowledge. Here too, for all the usefulness of this space, silences remain embedded in the soil and on the screen. These silences "[demand] a response" and I must "continue to speak where some would prefer silence" (Mackinlay, 2016a, 164). Embracing a renewed common garden story, also re-turn to reckon with a re-claiming discourse. The verge continues to contain an uncomfortable, colonising secret.

In Love & Fury

Renée

CHAPTER 4

Urban Food Street: A Vision of Community Connection

In the time before this time smothered lawn and hegemony across the land there was another time and place where the rich men wrote about her as if she, like the plants and animals in her garden, was something to be observed, noted down, written up. They watched but they didn't see. They forgot that the women who gardened the commons were ever vigilant. She observed them as much as she studied the slow, seasonal change of the common land.

She had no medium to tell her story, no way to write her heartlines (Mackinlay, 2016) *so her garden became the story, created with tools, and inscribed in the land. Her garden story, then, brought her mind in communion with other minds, now, even if she never knew it* (Le Guin, 2004, 209). *Community acts communicate but to do so they must all be playing the same tune. What lies beneath is where the stories lie, composted together. She left objects along the paths to lead the way.*

She used spades and shovels to dig up her dissent. Those tools remembered her, the way her hands smoothed their shape from over-use. Her mundane tools were disobedient. She used them to break down the hedges and the fences and to fill in the ditches they dug (McDonagh, 2019, 263). *Her tools recorded her life, but became blunt from over-use because it is on those tools that she lay the story of her life* (Ahmed, 2019, 22).

She senses their urgency. A change of use. They want to take the land that she has used in common and turn her garden use into land abuse. They made her an outlander of her own world with official documents buried deep in the bowels of bureaucracy. The pages are inked with words like council-owned,

© The Author(s), under exclusive license to Springer Nature
Switzerland AG 2024
R. Mickelburgh, *The Ecofeminist Storyteller*,
https://doi.org/10.1007/978-3-031-59242-3_4

69

state-owned, crown-owned but missing the words: people-owned. There is a story that can be remembered here. Their un-common sense of what could happen in their material world encouraged her to tell a different story.

They came and built hedges and the stone walls and fenced her off and fenced her in. She responded in defence of her communal life, situating herself at the "forefront of the struggle against land enclosure" (Federici, 2019, 107). *Her words and her resistance work were not compatible with their "fondness for straight lines"* (McDonagh & Daniels, 2012, 111). *Those straight lines sliced through her garden like a faultline. They unmade her world, turning her into an unwanted person. The fencelines were her unbecoming; they severed the heartline to her material world* (McDonagh, 2019, 262).

Her resistance to enclosure occurred underfoot as much as overland. Her feet opened the footpaths they closed; her tread made marks across the crops they grew instead of her garden. Tramping became her testimony. She laid down boards to cross the streams they had fenced her from and opened the hedgerows that closed her off and out. She inserted her body into their space and challenged their "property's grid" (McDonagh, 2019, 270). *Her memory was her gridline. It did not align with their property maps.*

Will they remember how she fought, how she in-sistered and re-sistered against the way they erected barriers and barricades with hedge-lines and fence-lines and forget-lines? She watches the wind pick up the soil and carry it over her gardens, smothering the landscape and memory of her life. The breeze blows her words into the distance, all but erasing her presence from history (Mortimer-Sandilands, 2008, 281). *Only the ones that listen closely will be able to hear her story now.*

When she pulled down the hedges, her body and her tools merged into one finely tuned instrument. Her body became the tool of dissent, in tune and in sync with the conversation. The material and the mundane and the body working together to resist threats to the common bonds and bounds of their communion (McDonagh, 2019, 262). *Getting to down to the root of the problem revealed a problem so monstrous they knew it would grow like a weed across land and across time. Her body and her things were both entangled in this encounter with enclosure.*

They say she inscribed resistance on the land, but the only pen she had were her arms and hands and feet. It was her call to arms (Ahmed, 2014, 194) *when their gardens were replaced with pastures for other animals to graze. She feared dissolving into the dust and so her inscriptions on the land were the words that would be pictured centuries later. She in-sistered (after Cixous, 2007) on these un-common garden rules to unmake what they made.*

She wonders what they will find once her footprints disappear deep into the soil like the roots of a tree. Will the history writers know where to find her story, that her footprints are her rootprints (Cixous & Calle-Gruber, 1997)? *How will they discover the way that she and her sisters tramped across their paths so her walkways remained free-ways? Their walking as speaking was as mundane as the flowers she grew in her garden. It was the common refrain of their bloom space. Reinhabiting is a word that inhabits many worlds.*

Her tools were a time traveller, travelling through time to tell the story of her use (Ahmed, 2019, 25). *They were not weapons of violence, but weapons of choice. They did not draw her enemy's blood, but dug up their fields. Her tools were a means of contending both against something and for something. They left a trace on the earth and along history's lines that revealed her opposition to the unfair advantage enclosure had over her world. Her tools inscribed on the land her dispossession from the soil* (McDonagh, 2019, 266). *Weapons aren't always violent, sometimes they are used for gardening.*

This chapter examines what happens when a community garden forged on suburban street verges becomes a space of resistance. The *Urban Food Street* project used suburban street verges in an Australian neighbourhood as the place people who lived side-by-side came together to grow food. This chapter considers how visual storytelling exposes something when something in the garden is dug up. I argue that when that something is the ordinary home space, traditionally a place of mastery and control, photographs of this un-earthing become plot points in a digital tale of transformation. Photographs become both a research tool, and a gardening tool when they are used to re-imagine a history of dissent and disobedience. Gardening tools, then, are both the photographs and the objects they depict. They become disobedient objects when groups of people join, en masse, to plough suburban streetscapes, grow shared suburban gardens, and then place images of those actions online. Photographs of shovels and pitchforks and wheelbarrows are re-turned, in this re-view, to their other historical uses: as objects of common dissent. This dissent is not violent, but rather it is breathtaking in its beauty. Visual imagery depicting a community re-turning *car*ways to *foot*paths, force a re-view, a re-turning, or a re-orientation to another time. Neighbours bring history alive in this physical and digital space through their "willful presence on the streets" (Ahmed, 2014, 163). By moving (and growing) together, they became part of movement away from convention. The *Urban Food Street* story became a place where bodies gathered together to "reclaim time and space" (Ahmed, 2014, 163). The community came together to re-work the thinking around the best use for a piece of land that they lived on. They made a deep acquaintance with this place. One women's pictures of this project grew a visual story that extended far from the boundaries of the neighbourhood it emerged from.

Urban Food Street was a high-risk resolution, food revolution where pixels stood tall amongst the radishes and bok-choi surged across the screen-scape. It was a suburban, urban spread where food plants were planted out of place and joy blooms in time with the seasons. Verge food became comfort food and visual stories of verge gardens signalled a beautiful, peace-loving way to live in communion, to create community. I wish this was the end of the story, but it is just the start. There is much more to the story. As the clothesline whirls, the pictures spinning around reveal it wasn't all smooth sailing. There are blank spaces between the pictures—gaps, spaces, silences—that punctuate like an irregular heartbeat. The story-line jumps and jolts and messes up the timeline and visual plot points in this narrative arc. Details are now missing, and so some things remain out of alignment, out of attunement. A visual boundary breakdown story broke-through the world of council control. By avoiding regulatory attention, it came to the attention of regulators.

The *Urban Food Street* project used suburban street verges in an Australian neighbourhood as the place people who lived side-by-side came together to grow food. When something is dug up, something is exposed. When that something is the ordinary home space, a place of mastery and control, photographs of this un-earthing become plot points in a digital tale of transformation. Viewing the images the *Urban Food Street*, first published in 2016, forced me to re-view what happens when property lines crumble and boundaries breakdown and stories, as well as soil, are exposed. If, as Stacy Alaimo suggests, "the home, the yard, the apartment complex, the gated community are places of mastery and careful demarcation of property lines—spaces of order and control" (2016, 20), then I wondered: what happens when images expose a story of un-control and dis-order? Exposure in this ordinary place has the potential to become an important environmental exposé.

Gardening tools—as both photographs and objects—become disobedient objects when groups of people join, en masse, to plough suburban streetscapes, grow shared suburban gardens, and then place images of those actions online. Photographs of shovels and pitchforks and wheelbarrows are re-turned, in this re-view, to their other historical uses: as objects of common dissent. Visual imagery depicting a community re-turning *car*ways to *foot*paths and neighbours walking and playing in streets meant for roads, force a re-view, a re-turning, or a re-orientation to another time. Neighbours bring history alive in this physical and digital space through their "willful presence on the streets" (Ahmed, 2014, 163). By moving together, they became part of movement away from convention. The *Urban Food Street* story became a place where bodies gathered together to "reclaim time and space" (Ahmed, 2014, 163). They re-worked thinking around the best use for a piece of land that a community lives on. One women's pictures of this project also grew a visual story that extended far from the boundaries of the neighbourhood it emerged from.

Urban Food Street used photographic images as digital plot points in its online narrative arc. They reminded those viewing the images (often far from this neighbourhood garden work) what happens when a community reorients a private, home garden space for common, public use. To consider the ways these images worked, is to consider what Gillian Rose says is the "trace of social identities, processes, practices, experiences and worlds" (2014, 33). *Urban Food Street* left markings that were visible far from the streets of Buderim. Their online presence makes them visible to a global audience. Such traces, Knowles and Sweetman suggest, have the capacity to "reveal what is hidden in the inner mechanisms of the ordinary and the taken for granted" (2004, 7). The visual activates the imagination by capturing "the particular, the local, the personal and the familiar while suggesting a bigger landscape" (p. 8). There was something about the photos I saw that left a trace. These markings saturated previously un-used, or under-used spaces with both bodies and gardens, the photos were the traces of this existence, left behind on the screen (Ahmed, 2014, 164). *Urban Food Street* images tell a story about what did happen, but also what could happen and what has happened locally. The digital publicity of these markings meant such a story emerged, and remains, in much more dis-tance places and times.

Something significant happens when images reveal an unearthing of the garden and unbounded the bounded space of the home (Alaimo, 2016, 20). I wanted to understand, as narrative analyst Catherine Riessman sug-gested, how and why these images were being used to communicate a narrative (2008, 142). But also, what exactly that narrative was, or per-haps, what it imagined life *could* be. Images are "visual representations of experience [enabling] others to see as a participant sees, and to feel" (Riessman, 2008, 142). I wondered what photographs "do" in visual nar-rative inquiry and what themes emerge from the visual digital story*telling* of *Urban Food Street*. There is the matter of materiality in photographs, even when they are in a digital medium. I felt visual narrative analysis would help me understand the way images both connect with and communicate objects. There is a sensuality to this un-earthing. When visual imagery is projected onto the screen and through the digital world "digital interactivity is configured on the technical and artefactual materiality that enables networks of communication and participation" (Martínez Luna, 2019, 46). I intersected this visuality with a semi-structured interviews with their creator, Caroline Kemp I searched for stories found in the images, but also compared and contrasted that with the photographer's intention. There was a sense of temporality interconnected with its mobility as a digital object. It disrupted a linear narrative, evoking memories of the past and remaining on the screen long after the photographic event has occurred.

In terms of affective communication, Ahmed provides a clearway to understanding what happens when we use things. I walked affectively with her along my path of visual analysis and to help explain use, and how to think of it in terms of communication, of how we make use of signs (2019, 49). These photographs showed how, when something undergoes a change-of-use on the ground, it also re-shapes what is communicated. When something ordinary is dug up—even something as simple as a street verge—something extraordinary happens. Ahmed extends on that thinking, revealing how, when something is used differently, something stops going with the flow (2019, 210). Stopping the flow creates an obstruction and an obstruction can also become a blockade. The *Urban Food Street* photos provided the frame through which to re-view use, to show how "frames of use have uses" (2019, 46), but also how using something differently is an effort in willfulness. When they dug up the verge on the edge of these ordinary streets and homes and exposed the soil to the world through visual imagery, the area no longer corresponded to its intended function: dominating the ground and growing lawn. They used it differently; they modified the form (Ahmed, 2019, 63). A neighbourhood converged on the verge and changed it, over time. It no longer corresponded to its intended function (Ahmed, 2019, 24). These communicative signs reached across suburban borders and boundaries, troubling the concept of use, common space and what it means to live in a community. They were on the verge of something.

The *Urban Food Street* images, projected from suburban front yards, onto the screen and towards a global audience, re-minded me to re-turn to Barad (2014) in my narrative journey; they re-turned me to all the *Re*'s. They re-turned the reader (or in this case the viewer) to the commons. They re-told a common story about re-claiming the commons. They re-turned to a narrative about what protest can be and what it has been when public land becomes not-so-public. They forced a re-consideration of the response-ability of re-turning to this commons discourse in the first place. Exposure, both of the soil and in the digital world, can be a secret revelation. But it also forces a re-view. A hope-full, but also fragile, optimism re-turns in *Urban Food Street*'s images. There is a fragility in a story about people re-turning to an ordinary place to unearth the soil that has been stolen. In Australia soil is fragile—both the substance and the story that surrounds it. Connections are easily broken. Accumulation and appropriation sit uneasily within commons rhetoric. Colonisation was the justification that something could be put to better use too.

BEING IN-COMMON

Urban Food Street, both the project and its digital imagery, grew from simple matter. Sunshine Coast social scientist and architect Caroline Kemp and her partner, retired horticulturalist Duncan McNaught lived in Buderim, a suburb on the Sunshine Coast in Queensland, Australia. Buderim has a multi-layered history. It is situated on the traditional lands of the Gubbi Gubbi people. European colonisation in the mid-1800s turned the area into farmland for sugar cane, and later bananas and citrus. In the late twentieth century it became renowned for ginger. Buderim Ginger remains a well-known product within Australia, although its famous manufacturing factory later moved to the nearby suburb of Yandina. Today the area is classed as a semi-rural suburban locality and much of the farmland has been broken up into traditional suburban blocks.

Kemp and McNaught's project began in the most simple, ordinary way. In 2009 they decided to grow some citrus trees. At the time individual limes cost two dollars each. They realised that once mature, the trees would produce more fruit than they could possibly eat, and they faced significant over-supply. The solution, they believed, was simple: plant lime saplings on the verge in front of their property so others walking past might be able to pick some fruit. In our interview, Kemp explained that their lime-tree plan expanded after she and McNaught discussed their verge-planting plan with a neighbour. The neighbour liked the idea and wanted to do the same thing. McNaught's connections in the horticulture industry allowed him to purchase good quality stock at wholesale prices and so, they bought eighteen trees for their own nature strip, and an additional six for their neighbour. Over time other neighbours became interested and so the idea grew, over several years, into a project that eventually encompassed twelve neighbouring streets.

There are no official figures on the total number of homes or people involved in *Urban Food Street*. Kemp estimated that it may have encompassed up to 1000 residents. The project grew informally and there was little sense of hierarchy or overarching management or obligation. Just like the produce its residents harvested, *Urban Food Street* developed in an organic, relatively unstructured, and relaxed fashion. There was no official committee and only a few meetings convened in the early stages of the project. Kemp said that there were "some minutes floating around somewhere". In line with the informal structure of the project, its expansion from the front of two properties into twelve streets did not require or utilise any formal recruitment process. Neighbours simply saw (and tasted) the benefits of a local, free food project. A few years after planting their lime stock the trees reached productivity and Kemp and McNaught invited neighbours to pick the fruit. After seeing (and tasting) the benefits of local free produce growing on front verges, neighbours approached Kemp and McNaught, keen to start their own verge gardens. Discussions ensued about the residents' time restraints and other personal responsibilities, their growing knowledge and ability, and what their site could accommodate.

Urban Food Street brought the home to the street. Not everyone in the 12-street area physically participated in gardening. There were many reasons for this: some had extensive caring responsibilities for young children and babies; some were elderly, or physically unable; others just didn't have the time, or interest, in growing on their verge. Yet despite not wanting or being able to physically create gardens, many were keen to contribute to the project in other ways. For example, one neighbour might contribute their hose and water, so another neighbour's verge garden could be watered. Others would donate home-made compost or attend BBQ fundraisers. The project stressed flexibility to ensure commitment remained within a household or family's capacity. It was an opt-in or opt-out arrangement. The one important and unwavering rule of *Urban Food Street* was that everyone in the neighbourhood could pick the food, no matter what their contribution to its growth. This acknowledged the verge area as public space; the produce had to be available for everyone. As people's lives and responsibilities changed from year to year so did their involvement. For example, a new baby in the family may limit a household's capacity to contribute for a period. As the project expanded and more people became involved weekend working bees and fundraising BBQs were organised. While many houses were not directly involved in planting on their verges, most attended social events. The unique skills and abilities of those who lived in the area contributed to the success of fundraising BBQs. For example, one resident was a chef, and so donated coffee and food, allowing all money raised to go straight back into the project.

Communicating their project initially occurred in a very grassroots, local fashion. There was little that was digital about it. A chalkboard notified nearby residents of working bees and fundraising barbeques. Neighbours created a monthly, paper-based newsletter which they dropped into letterboxes throughout the neighbourhood. Kemp initially resisted any sort of digital or social media presence. She said she and McNaught saw community as "ill-defined" whereas neighbourhood gave them "a target audience and a boundary." They wanted to keep communication in this neighbourly sphere, rather than what they saw as the wider, harder to measure community. She says the distinction between the two was important.

The social-technical communication process started when Kemp began recording the story of the lime trees through words and photos. She initially placed these in an online blog and then gravitated to social media sites. Digital photographs were a significant part of the project's online story. The project had a strong visual presence on Facebook and Instagram, as well as its own website. This photographic imagery, and the online medium, meant the *Urban Food Street* story was able to travel well beyond the boundaries of its suburban neighbourhood. Its presence within digital platforms allowed it to reach a global audience. *Urban Food Street* did not start, or even continue, with a specific online communications strategy. Kemp would post when it "felt like a good idea". Sometimes she posted daily, sometimes not for several days.

A Common Change of Use

Like most Australian verges, *Urban Food Street*'s activities centred on a space where lawn was traditionally grown. The digital images prompt the question: is this the best we can do with this space? Ahmed says, "we learn about something by considering how it is being used, has been used, or can be used (2019, 22). The *Urban Food Street* images revealed alternative uses for the space that had been used for lawn, a hegemonic monoculture which dictates middle-class Western views of private property and the home (Robbins, 2007; Robbins & Sharp, 2003). Lawn discourse embraces certain middle-class cultural and economic imperatives. It is represented as the "normal" aesthetic: a sign of property ownership and citizenship. Growing and maintaining lawns is an environmentally detrimental practice. It is not useful to native plants or animals, nor does it feed or provide water or shelter to human life. Maintaining its green, well-manicured standard is a time-intensive process that requires suburban citizens to use environmentally hazardous chemicals. The lawn is a space where, "citizenship, ecological metabolism, chemical hazards and economic imperatives come together every-time someone practices intensive lawn care" (Robbins, 2007, 96). In Australia, the verge is usually lawn, and mostly council-owned land (this varies depending on the state and local government area). It is ostensibly public land, that the householder is expected to maintain.

In the *Urban Food Street* imagery, the traditional use of the verge—as a place where lawn had been grown—was slowly replaced with stories of how it can be used. Lawn slowly disappears and soil emerges. As the pictures progressed over time, small food gardens are planted and produce blooms and flourish. Freed from its lawn-sustaining burden, images of the exposed soil offered the potential for a different use, and so pose a different story. During our interview, Kemp revealed the project evolved from a few lime trees towards an intention to put the verge space "to better use". As she spoke, I remembered Ahmed's words, that "things are transformed by being useful" (2019, 25). The images show the lawn verge as a monocultural thing being transformed into something more useful. The exposure of soil suggests the potential for better use of this space. Turning the verge into a shared habitat hints that planting is more use-full than mowing (Fig. 4.1).

Changing use on the street and reflecting that on the screen encourages a broader re-view. What else could be put to better use? Is any of suburbia useful? Lawn dug up but the surroundings—contemporary suburban life—remains the same. In the mediated pictures of *Urban Food Street* contemporary suburban life remains present, but backgrounded. Suburban houses revisible, side by side. Ornamental trees and shrubs, cement footpaths, modern family cars sit in driveways and bitumen roadways clearly write the story of middle-class Australian suburbia. These are the goods

Fig. 4.1 Creating verge gardens. (Source: From *Urban Food Street*, 2016e, https://www.facebook.com/urbanfoodstreet/photos/571702993009905). Reprinted with permission

and structures of the dominant, consumer-driven economy, embodying the many elements of what constitutes the "good life". Pictures of disturbed lawn and exposed soil unsettles the streetscape, and so it also disturbs those things that frame the lawn. A lawn borders a house, it rubs up against fenceline. When there is no lawn to act as a border, the border begins to fade. Disturbing the soil also unsettles common-*sense*. Lawn, then, becomes un-common sense. It is a garden product which has long smothered secrets that are now being revealed by digging up the space and exposing the iron-rich red soil.

Willfulness merges with usefulness as the *Urban Food Street* stories emerge on the screen. Reinhabiting the verge space is also a reorientation of that space. The residents don't wave placards or chant a chorus of protest, but they do assemble on the streets in an unusual way. The residents form an assembly line to change the use of the verge and re-turn to Ahmed's words about the way assembling on the streets "can be a protest against how and by whom the streets have been owned" (2014, 163). There is a sense of disobedience in these pictures. The images do not depict violence or anger, but they show a group failing to conform with the "lawn and order" (Butler-Bowdon, 2001; Sandberg & Foster, 2005). As they assemble the images story*tell* a design *experiment*-in-place, because "objects that are available to use do not allow us to do something as well as they could" (Ahmed, 2019, 25). The story: the verge was not doing as well as it could. Digging up the lawn—modifying the form—brought something new into existence (Ahmed, 2019, 25). The form transformed into something more useful (p. 25). The verge re-emerged as a shared habitat of suburban humanity, ecology, and technology. Ahmed says modifications can also be reorientations (2019, 63). The imagery reveals what happens when there is an opportunity to grow something new and productive.

Mass Ploughings

Images of suburban verges being dug up carve a line towards historical memories. This is a sense of mass ploughing in the *Urban Food Street* imagery. Reorienting the use of this space draws a temporal, material, and physical line towards "mass ploughing", resistance methods common people used to fight English commons enclosure. Mass ploughings occurred to restore the English common landscape to a state of pre-enclosure by "visibly and materially resisted enclosure by returning the land to its pre-enclosure state" (McDonagh, 2013, 2019, 265). This "visual marker of dissent that was visible far beyond the boundaries of the close: the plough literally inscribed resistance into the surface of the land" (McDonagh, 2019, 266). In the *Urban Food Street* images a home ground is being re-shaped into a more useful form. It is a form that will benefit a larger group, not simply an individual. There is a hint of a new common-sense where the home ground is being viewed from a different perspective.

As the visual soil story expands, so does the human action. Human presence is required to dig up the soil for a change-of-use. Soil patches become larger; and more human power is required. Photographs depict neighbours busy working the soil, others are watching, some are talking. They are people of all ages, from young children through to adults. Men, women, and children are pictured, participating in these digging projects.

Kemp stated in her interview communication, that *Urban Food Street* did not set out to reclaim the commons. She says, "It was very innocent, it was literally influenced by the price of a lime and wanting to have a more active street really. Were we thinking of the commons at the time? No". While it may not have been a conscious decision, nor even part of their written narrative, the imagery—the storyshowing as storytelling—re-turns the viewer to a sense that reclamation is occurring. Reflecting on the project, Kemp said during our interview that *Urban Food Street* did end up becoming a tale of commons reclamation.

> If you look at it in its raw form, we're reclaiming public space and not being dictated in how we will use it. It wasn't a conscious, kind of we're going to reclaim public space it was like, yeah, I guess it depends on how you want to look at it. We wanted to activate. So maybe that in a sense, in itself is a reclaiming of the space by saying we need to activate this and get it more people focussed. (C. Kemp, personal communication, June 6, 2019)

The more pictures are uploaded and circulated the more the uncommon form—a food-producing verge garden—becomes common. As the images progress, more of the lawns vanish and exposed soil changes form once again, morphing into garden beds. The pictures reveal what is being re-shaped in the space: a garden. The change of land use is reflected in the pictures. What had been in the frame was removed and what had previously been out of the frame, re-turned. There is a sense of being or becoming in common in this frame (Gibson-Graham, 2006, 86) as the physical, social, and technical mingle and merge. The digital, visual narrative becomes an evolving, but also revolving, story. It is a change of use story: suburban lawn is re-turning to farm, if only on an urban scale. Lawn is replaced with soil, which is then replaced with small green, leafy seedlings in defined rows. Garden beds emerge and are planted with herbs, a simple gesture which helps residents to avoid the sometimes one-and-a-half hour return trip to the closest supermarket. They are now able to wander out on to the street to collect what they needed. The verge becomes a "scene of breakage", where something that had been un-used is put back into prior use (Ahmed, 2019, 21). This is a useful transformation (Ahmed, 2019, 25). Of course, this prior use only extends so far. It remembers previous agricultural history, but not an Indigenous history that extends far further.

Instructions can be restrictions and restrictions rely on enforcement. *Urban Food Street* images parallel the historical mass ploughings of commons enclosure by suggesting a "highly visible symbol of an enclosing landlord's lack of authority in a locality" (McDonagh, 2013, 49). Most of

Fig. 4.2 Crops replace lawn on the verge. (Source: From *Urban Food Street*, 2016c, https://www.facebook.com/urbanfoodstreet/photos/534042480109290). Reprinted with permission

the time the project did not defy council orders, it simply did not engage with the local authority. But not asking permission is also a way of contesting authority. It is not that *Urban Food Street* overtly objected to restrictions. It is just that the instructions were not clear, and so they made their own (Fig. 4.2).

Disobedient Objects

In *Urban Food Street* photographs transform the mundane, everyday domestic gardening tools into objects of disobedience. This re-view does not reveal them as weapons of violence, but rather tools that disrupt the status quo because of what they are being used for. In doing so, this digital imagery mingles with a historical legacy of the way these same tools protested commons enclosure. In the mass ploughings of English commons resistance, mundane and everyday objects were "entangled" in

Fig. 4.3   Gardening tools: mundane or disruptive? (Source: From *Urban Food Street*, 2016f, https://www.facebook.com/urbanfoodstreet/photos/571714109675460). Reprinted with permission

anti-enclosure rioting (McDonagh, 2019, 254). Tools were used to dig up fields and pull-down hedges were part of the resistance. They were out of place within property boundaries, becoming "riotous, disorderly or otherwise disruptive of existing or new agricultural arrangements" (McDonagh, 2019, 271). There is a sense, within these pictures, of Ahmed's words about objects used in the past, pointing us towards the future (2019, 22). Despite the availability of more sophisticated technology, the tools pictured in *Urban Food Street* re-turn the viewer to those entangled with commons enclosure and the resistance that was written through tools rather than pen (Fig. 4.3).

Remembering Ahmed's insistence on learning about something by considering how it is being, has been and can be used (2019, 22) then the tools used in *Urban Food Street*—the ordinary, everyday objects—contain oppositional meaning. If "use records where we have been, use can also direct us along, certain paths" (Ahmed, 2019, 23). Travelling from the

Fig. 4.4 Gardening tools recall another era. (Source: From *Urban Food Street*, 2016g, https://www.facebook.com/urbanfoodstreet/photos/ 1307231679457029). Reprinted with permission

path of enclosure resistance to the photographed streets of *Urban Food Street* shows how tools here too are used to dig up what, to residents, had become use-less land. There is a scarcity of modern high-tech machinery throughout the stories. Such machinery is easily available and would have made their work faster and more efficient. Instead, the garden construction relies entirely on the human body and mostly manual tools such as shovels, pitchforks, and wheelbarrows. Past, present, and future use mingle in the garden tool imagery. They are forms that transform both the physical and digital space. The physical is dug up using these tools, but so too is the way dissent can be communicated (Fig. 4.4).

Roads as Common Space

A path's becoming occurs when a custom becomes customary. Roads became walking paths when they bound gardens. This becoming-a-path change of use is fuelled by feet. Feet are what re-turn us to forgotten customs of past use. *Urban Food Street* embedded custom through the comings and goings of bodies and made the use of streets as walking paths customary (Ahmed, 2019, 41). They left disruptive digital traces in mundane, everyday online place (Fig. 4.5).

As the *Urban Food Street* images drift from the verge onto the bitumen, they contest what is common-sense, questioning the use of a road and the use of a path. These images become a story about bodies moving where they could before, moving in areas not just where they *should* be, or ought to be. Pictures of roads for cars become paths for feet. The digital imagery resists the roadway as a car-way story and instead tailors this space as a

Fig. 4.5   Human traffic. (Source: From *Urban Food Street*, 2016d, https://www.facebook.com/urbanfoodstreet/photos/567795560067315).   Reprinted with permission

place of re-use—a space for human traffic. Pedestrians and domestic animals and non-mechanical modes of transport such as bicycles and scooters take up a space which is usually allocated for vehicular use. Bodies take up space across the road, rather than existing on its edges. Adults and children congregate for impromptu games. There is a queerness to this story, the pictorial narrative is "odd, noticeable or curious when it reverses how things exist usually or by challenging how things are expected to be" (Ahmed, 2019, 75). In Australia, roads are an area mostly exclusionary to pedestrians. Unlike countries such as The Netherlands, for example, these streets are not designed with cyclists or pedestrians front of mind. Pedestrians are used to sticking to footpaths (if there are any), and cyclists to the edge of the roadway (Fig. 4.6).

The visual *Urban Food Street* story re-turns the roadway to the people-way. This imagery reveals their moment of being. Photos that re-turn neighbours to the road are also re-turning to a historical moment of

Fig. 4.6 Roadways as common ground. (Source: *Urban Food Street*, 2016b, https://www.facebook.com/urbanfoodstreet/photos/487343611445844). Reprinted with permission

movement, when movement through space becomes a "willful presence on the streets" (Ahmed, 2014, 163). Historically in early modern Britain, the right to move through designated pathways were negotiated through "local custom and the concept of neighbourliness" (Whyte, 2015, 394). Roadways as common space became "negotiated spaces" where movement through them—choosing paths and meeting people— "were political in the sense that these most basic requirements of everyday life were determined and controlled by social and cultural expectations and conventions" (Whyte, 2015, 934). Pathways connected people, transported goods, and shaped both the landscape and people's interactions. They also highlighted the non-linear shape of enclosure; it was negotiated depending on seasons and community needs (p. 929). Pathways determined the social interaction that occurred in a community. Environmental and climate change issues were not the concern of those living in this period: the roadways were used for basic survival. Yet both contemporary use depicted in the *Urban Food Street* photos, and the corresponding historical use, was willful. It became an obstacle to the way property flowed. Movement along these paths re-ignited flow that had been cut off at some stage— either by hedges, or bitumen roads.

This human visual digital story of bodies on streets is as much a part of the narrative as the growing of food on the verge. It was a conscious, proactive decision to post these images of bodies in the street, Kemp said, in our interview. The paths recall and highlight the historical use of the area: a lack of footpaths means people have often walked on the streets. Pedestrian foot traffic along the road is unusual and slightly absurd in the twenty-first century; images of bodies walking along these spaces are filled with contradiction and nuance. The space these roads occupy is being shaped by the feet that tread there, rather than cars that drive there (Ahmed, 2014, 163). The *Urban Food Street* imagery resists conventional road use in a very a specific, localised way. Kemp admitted that representations of this human action on the road were easier to produce than a similar action occurring on a busier roadway "like (Brisbane's) Coronation Drive (where) there's 4000 cars going down every hour". The *Urban Food Street* roadways, in contrast, only saw around 30 car movements a day and were relatively quiet in comparison. Ahmed says, "a history can come alive as a willful presence on the streets" (2014, 163). The bodies gathering along the roads in *Urban Food Street* are also bodies reclaiming time and space.

In *Urban Food Street*, like most of urban and sub-urban Australia, the bitumen road (as well as fence lines) creates a grid around and across the neighbourhood. English commons were enclosed property grids created for the first time through the establishments of hedges. Much like the difficulty entailed in removing hedges as part of enclosure resistance, bitumen roadways are also relatively immovable structures. Similar to the historic English commons resistance, those walking or using the roadways in *Urban Food Street* do so in the daylight hours, their bodies temporarily occupying enclosed space. There are no placards on the street; nor are there marches in protest. Yet the bodies in *Urban Food Street* question "how and by whom the streets have been owned" (Ahmed, 2014, 162). Of course, *Urban Food Street* residents are not using the roadway to access land for survival. It is doubtful that any will be forced into a life of abject poverty and despair as those who fell foul of English enclosure laws did. However, that does not mean that imagery of bodies assembling on roadways is unimportant or trivial. The way something is used suggests how it should be used.

A community can put a sign to a better use, using it for the very thing it was meant to be against. A sign can be used to communicate movement toward something even if its intention was to communicate movement away. *Urban Food Street* put detour signs to better use. If, as Ahmed says, usability can be treated as communication then a sign can suggest "whether it is obvious how to use something but (also) who can something" (Ahmed, 2019, 59). These signs, created and erected to communicate a detour against something were pictured and captioned detouring towards a radical a departure from the designated path.

Urban Food Street photographs include a sign directing feet away from something to encourage a detour towards something. The *Urban Food Street* verges were about six metres wide and included a swale, or depression, near the roadway. Many streets within the project lacked artificially created channel and curbing—there was no boundary between the edge of the bitumen, the swale and the rest of the verge. This allowed rainwater to travel from the roadway onto the gardens, freely nourishing the plants. Rather than complain about the lack of road infrastructure *Urban Food Street* welcomed its scarcity, pointing to the benefits of natural water flow. Kemp specifically highlighted the benefit of the swale in many of the pictures. A detour sign, urging people to detour away, encouraged them to a detour toward a place where the use-less waste of rainwater was contested.

Rainwater is useful to the garden verge; it waters the gardens rather than being channelled uselessly down storm water drains and into the ocean, taking with it plastics and other pollutants. The visual *Urban Food Street* story had digital plot points pointing which pointed out the ways free water should be water-in-common and put to use nourishing food producing plants. One *Urban Food Street* caption read:

> We loved the irony of this sign which was erected on the Mediterranean Cnr. during some unwanted and unwarranted road works in our area last year. The detour sending people off the beaten track and through our food abundant streets, effectively providing exposure to a slow and greener way of suburban occupation. Happy detouring! (*Urban Food Street*, 2016a)

The queer usage of the detour signage mingles with historical usage of space and complaint, transforming the direction of the detour. Again, a temporal messiness re-turns to this digitally visual tale. In English enclosure protest, resistors cut banks to re-establish water into newly drained land. The cuts ensured that "as the liquid transgressed the drainers' banks and ditches and leaked back onto the land, it remade the landscape—or perhaps more accurately, re-established a waterscape—thus resisting the imposition of private property relations" (McDonagh, 2019, 271). *Urban Food Street* used the detour sign to encourage others to make a detour on their thinking about the way water is detoured. In doing so they resisted the enclosure of water, and its wastefulness, by reiterating its common usefulness (Fig. 4.7).

Fig. 4.7 Detouring towards something rather than away. (Source: *Urban Food Street*,   2016a,   https://www.facebook.com/urbanfoodstreet/photos/a.395625327284340/444088652438007/). Reprinted with permission

The Story of a Community Economy: Joy, and Hope as a Story of Resistance or Cruel Optimism?

Happiness is digitally mediated in the human faces of the *Urban Food Street* story. Happy bodies signal using something differently generates positive feelings like joy and pleasure. Happiness suggests that using something differently generates images of happiness, but it also shapes the landscape. Something is working better than it has been. It shows that "natural environments are also social environments" (Ahmed, 2019, 63). Susan Griffin says, "at the core of human imaginings (is) the desire to locate ourselves in community" (2015, 180). If, as she says, "the wish for communion exists in the body" (p. 180), then bodies become the boughs of branches that reach for happiness in this gardening story. Here are the hands that dig up the lawn. Happiness might be about being in community, but it is also about being in communion.

The *Urban Food Street* pictures pleasure. In one picture McNaught stands, laughing, as he holds a young child up to pick bananas. It is unclear exactly what is bringing McNaught joy, but it clearly has something to do with the large bunch of bananas they are looking at. Perhaps it is the sun shining through in the background, perhaps it is the determination on the face of the young child, trying to pull down a bunch of bananas that is almost bigger than his entire body rather than simply grabbing some from the supermarket shelves. Whatever the true reason for the joy and happiness of man and child, there *is* joy and pleasure emerging from the transformed verge form. It can be seen in McNaught's wide smiling face but also in the juxtaposition of the young child clutching a bunch of bananas that is almost bigger than their whole body. In another photo an older woman stands next to McNaught within the transformed verge form. They are in what appears, within the frame of the photo, to be a lush garden. She holds a walking cane, and reaches out to him, smiling. The viewer cannot know what they are discussing but their smiles are a mutually understood signal for happiness and joy. They seem to be at ease with one another in this space, shaped by mass ploughing and disobedient tools. The space is the place their share smiles and what appears to be laughter. Joy has emerged from this re-used space. Similarly, material objects of appreciation emerge from this re-used space. Another picture shows a thank you note Flowing script from the letter flows from a body that has taken care and time to write by hand. It takes time, thought and resources to create and send such a letter. The letter would not have existed without the change of use that has occurred on the verge. Perhaps pleasure here is emerges from nostalgia. Disenchantment with economic imperatives is often expressed through nostalgia for the material objects and places that do not figure regularly in everyday life. The letters suggest people in the neighbourhood derive pleasure-in-common from this transformed place.

The absence of abundant pleasure in broader ecological discourses suggests pleasure has no ecological use. This pleasure deficiency suggests such an emotion is "downright opposed to ecological principals" (Sandilands, 1999, as cited in Alaimo, 2016, 27). Yet an absence of joy within environmentalism is problematic because, as history shows, reasoned argument is not enough to sustain social action (Johnston, 2008, 103) but also because, as Stacy Alaimo notes, the appeal of joylessness within ecological discourse is "hardly appealing" (2016, 27). There is a growing argument for the resumption of joy into environmental communication. Alaimo insists an ethics of inhabiting should emphasise "possibilities for a multitude of sustainable pleasures" (2016, 30); Josee Johnston says "positive, pleasurable alternatives to consumer seductions are required" (2008); and ecophilosopher Kate Soper urges us to reconsider "the good life" to include "the sensual pleasures of consuming differently (2007, 211). Perhaps, then, pleasure, joy, and happiness, then, are useful. If seeking happiness within the "good life" relies on the ecologically devastating practices of consumer dissatisfaction and never-ending consumption then a story that rejects such a discourse, but still engages with joy, pleasure and hope hints at an alternative, transformative mode of environmental communication. Pulling pleasure and joy out of dis-use requires making it useful. Placing images of pleasure and happiness onto the screen to tell a story about connection and community, suggests these emotions have an important place in any story of environmental protection.

A blockage can divert something, but diverting something can also to enable a more useful flow. Much like the detour sign, the bollards pictured in *Urban Food Street* imagery change the flow of traffic so that a space might be put to better use. The bollards block traffic, but in doing so allow pleasure to flow. Bollards clear space for play and celebration, for the communion of bodies rather than vehicles. These playful images of blockades communicate the flow of pleasure. Photos of hay bales placed across the roadways close streets to traffic for an annual Christmas street party. In our interview Kemp revealed that up to 400 people would attend the parties; some were "external to the neighbourhood". She says,

So, we'd have these big street Christmas parties and they were huge and we just took it upon ourselves as a neighbourhood to just close down a few streets. If you look at our urban formats they are back streets, no one uses them, we are closing them down to our own internal traffic for a few hours and that's all. It doesn't affect anyone else. So, we would just put hay bales across the road and shut them down and the kids could spew out everywhere and no one had to worry. (C. Kemp, personal communication, June 7, 2019)

The sense of "we" in the visual *Urban Food Street* story suggest something more than gardens is emerging on the verge. Pleasurable practices that generate kinship between humans, but also between humans and the more than human world are blooming alongside the produce. This sense of "we" is a connection, it is a willful spark that is lighting up connections not only on the ground but through the digital. It is a pictorial re-view reminds me of the spark between Wright and MacArthur and all that emerged from writing of the joy (and at times despair) of their gardens and their sense of connection to the natural world. If, as Ahmed says, proximity is what you are fighting for and separation what you are fighting against (Ahmed, 2017, 82) then dwelling in places of pleasure is also dwelling in "abundantly inhabited places of transformation" (Alaimo, 2016, 38). Inhabiting places where pleasurable interconnection takes place, sustains, and entangles human and more than human relations. When boundaries between home and neighbourhood fade with the destruction of lawn so too does the bounded space of the home, a space where the human is protected from the outdoors. Happiness is entangled with hope in this pleasure story. The joy of the unexpected "embraces possibilities of becoming in relation to a radical otherness that has been known as 'nature'" (Alaimo, 2016, 18). There is hope in the garden, but it is fleeting. There is hope in the digital imagery, which still remains on the screen. Images of hope emerge in the faces of both the planters and the players, that food will be grown, and meals will be had and shared. This sense of hope, that something is becoming in common (Gibson-Graham, 2006) emerges in flowers and smiling children and thank you cards. Expressing hope for another, radically different world, "is a political action" even when faced with despair and exhaustion (Ahmed, 2004, 186).

Hope and happiness can also be a cautionary tale when it contains a digital imagery generates intimacy mixed with privilege. Emphasising hope can over-estimate the ability of individual will and happiness too "can look like the face of privilege" (Ahmed, 2010b, 11). The home is a domestic space which was, for the most part "constructed to contain (privileged) women" (Alaimo, 2016, 18). Ahmed urges us to question the happiness we defend to recognise possibilities "to wonder about the present by wondering about the how of its arrival. What happens with the familiar recedes" (Ahmed, 2010b, 218)? *Urban Food Street* was gardening for change that relied on individual work in a relatively privileged situation. Such moments, which can also make us "blind to, or overwhelmed by, systemic inequalities" (Johnston & Cairns, 2012, 229). The *Urban Food Street* images may make viewers pause, they may charge us with willfulness and usefulness, they may allow us to re-view the wonders of pleasure and joy, but does this re-view extend to the analysis of privilege inherent in a domestic garden story?

THE RISK AND FRAGILITY OF RE-OCCUPYING THE COMMONS

Avoiding absences requires turning away from places where unsettling things happen and have happened. Perhaps silence, then, is what is cruelly optimistic within the visual *Urban Food Street* story. It is a project that I, and many others desire, but its silences remained an obstacle to its flourishing. To keep what is absent in this story a secret would be to "preserve the possibility of (my) own research" (Ahmed, 2010a, xvii). Silence also avoids the discomfort of digging further down to explore the complicity in my own existence as a gardener and home-owner and technology user in Australia. The silence and absence in this place of modern online-offline roadway resistance is a secret that needs to be communicated in any contemporary feminist research about a physical and digital reclaiming of the commons. Theory becomes porous when convenient forgetting sustains utopian garden analysis. The truth is stories about Australian places often contain missing pieces and so the picture remains incomplete without them. Sara Ahmed says at times "we might need to reveal a secret to refuse a social bond or challenge the power that controls information" (2010a, xvii). And so, as much as I love the story I tell about *Urban Food Street*, I also try to dig up this secret that has been covered in rituals of habitation, from now and then, where colonising ways of gardens changed the meanings of landscape and place in this country (Holmes, 1999, 152).

Australia's common wealth relies on the dispossession of Indigenous people from their cultural lands. This common wealth includes the growing of gardens. Gardens are grown on stolen land, whether within the bounds of a fenceline, but also even when those boundaries begin to be broken down. The unspoken risk to the visual *Urban Food Street* story is that this type of project is simply the re-telling of another colonial garden story—that British idea of ritually habituating stolen lands (Holmes, 1999, 152). Berlant points out that while those proclaiming the common's "'manifestic function' is always political and invested in counter-sovereignty", the reality is the notion is still performative and aspirational (2016, 397). Those reclaiming the commons concept contest ownership rights and resource justice through performative acts. The Occupy movement is one such example. In doing so they "tap into legacies of occupation to contest ownership rights and resource justice", however, under the commons concept communities also project "a pastoral social relation of mutual attachment, dependence or vitality" (p. 397). The complex reality of the commons concept is that what is common to some is not common for others. Tuck and Yang's refrain that "decolonization is not a metaphor" (2012) re-emerges in commons discourse and here too it emerges in the *Urban Food Street* re-view. Any suggestion of claiming land of the common use, say Tuck and Yang, "erases existing, prior and future Native land rights, decolonial leadership and forms of self-government" (2012, 28). Moreton-Robinson says that what is in common for Indigenous people—dislocation, as a result of land being acquired for new immigrants—occurs alongside what is in common with migrants, the shared benefits of Indigenous dispossession (2015, 17). For the garden to hold a different value, Jennifer Hamilton says, "requires nothing less than the complete restructuring of the dominant land holding system" (2019, 483). *Urban Food Street's* visual imagery depicts an intention was to put unused or under-used to better use. While there is much achieved in this endeavour, there is still much to do.

It is important to note that from the outset Kemp and McNaught were reluctant to describe their project as political. They never set out to make a statement about reclaiming the commons, nor have it politicised. There is little in their publicity of the project that insists on a re-turn to the commons. In part, this was purposeful. The pair believed neighbours would be more receptive to an apolitical project. They didn't want their actions to be seen as activism, or radical. In our interview Kemp said:

> We wanted it to be seen as an alternative way that provided something for the local neighbourhood, for the people that lived in those streets. We wanted it to be seen as a project that enabled people to have an ownership

over where they lived and to use the streets in a way that suited them. (C. Kemp, personal communication, June 7, 2019)

Despite their apolitical intent, any public space in Australia contains colonial privilege and therefore is inherently political. Whether they liked it or not *Urban Food Street* was political because "public spaces are a political issue" (Apunkt Schneider & Friesinger, 2010, 17). People communicate through public spaces not only through daily interactions with each other, but also related texts and institutions. Public spaces are themselves texts through which control is exerted by those in positions of privilege (p. 17). A public space is a site of "an articulation of power relations, an ensemble of hegemonic symbols, in which dominance decrees itself and its presence" (p. 15). In Australia land is saturated in colonial silence and so it is always political. Even when a story makes no claims about reclaiming the commons, when it is a story about putting land to better use for cultivation and community, in Australia public space remains a text with inbuilt silences.

Whether it is using the verge to grow lawn or putting the space to better use for urban food production, gardening on the verge remained a colonial act of gardening on invaded land (Holmes et al., 2008). When use is defined as cultivation it risks a colonial refrain. The colonial project has long justified itself for making better use of the land, "as using what is unused" (Ahmed, 2019, 47). Hamilton argues for a new type of land system on which the garden sits; one that is "anticolonial and multispecies" (2019, 482). A garden sits on a land system which "legally underwrites colonial invasion and thus enables the ongoing indigenous apocalypse" (2019, 482). Hamilton urges for a botanical reckoning with this complicity in the colonial project, pointing to the need "to burn the picket fence and open the front gate" (2019, 483). *Urban Food Street* went part of that way. Its visual imagery did depict neighbours (and strangers) wandering out their front gate. There were no flames, but lawns were dug up, boundaries blurred, and objects objected to the status quo. These images of a new type of garden dissolved fencelines and boundary lines. And yet, doing exposed *Urban Food Street*'s fragility. The failure of traditional form was the reason behind its success, but also the cause of its sudden and brutal demise.

Much like the growth and demise of a tree, the visual stories of *Urban Food Street* took years to develop and only minutes to destroy. In May 2017, Sunshine Coast Regional Council destroyed the verge trees. Kemp and McNaught were not present on the day of the destruction. Neighbours recorded vision of the council officers destroying the trees. Kemp uploaded the content onto the *Urban Food Street* Facebook site. There is no speaking on the footage, just vision of trees being placed in a woodchipper, and the sound of the trees being destroyed. When questioned about the implications of the footage, Kemp says her decision to place someone else's footage on her website makes a statement about a project, that was "delivering so much good" also shows "government is willing at all costs to do this action to it". At the time of analysis *Urban Food Street* no longer existed in the ways these visual images depict. Kemp told me that along with the citrus and banana trees, the sense of neighbourhood, the working bees, community BBQs and Christmas celebrations all disappeared. Much of the verge has reverted to grass, which often isn't maintained. During our interview Kemp noted that the neighbourhood streetways are now very degraded, and people no longer congregate on the verge. She said that while neighbours still get together, it is often inside, "as if the street is the forbidden ground". There is fragility in the *Urban Food Street* story. After seven years of operation, the project was destroyed in a matter of hours.

Kemp posted a video of the destruction on *Urban Food Street*'s social media accounts. What was breakable, broke. I discussed this event with Kemp, who felt strongly about the reasons behind the destruction (C. Kempt, personal communication, June 7, 2019). Part of that conversation is as follows:

Caroline   It was a raw fact that here's a project that's delivering so much good. We can't actually locate another project Australia wide in this realm that's delivered so much. We can't even locate one internationally and we've looked hard. And government is willing at all costs to do this action to it. Because they perceive there's a liability problem. And so, the video was kind of raw fact to us. And so, we were like right, this is what they will do, they'll stop at nothing.

Renee   Why did you feel it was essential that you posted it?

Caroline   The video? Because I think people needed to know what government will do when you're not being obedient to them. It was a matter of obedience, right? We weren't applying for a permit because we knew this whole permit stuff would stifle what was happening and would destroy kind of the growth or

> potential for other people grow a project like this. What's
> interesting in that is that we now have some rules on the sun-
> shine coast and yet you don't walk around finding people
> planting verge gardens everywhere.

The project's sudden demise was what it always risked. The features and strengths depicted in the *Urban Food Street* imagery also signalled its fragility. The narrative thread linking the story—its lack of authority or respect for authority; its joyfulness, pleasure, and happiness; and the colonial legacy of land title in Australia—were also what made it a precarious story. Kempt told me in our 2019 interview that in November 2016 *Urban Food Street* invited the local mayor and the local councillor responsible for town planning to its annual Christmas celebration. They declined. Three days after their Christmas party *Urban Food Street* residents received a letter from the Sunshine Coast Regional Council informing them that their actions in relation to *Urban Food Street* were illegal, and that they were using the "road reserve in a way that was against council bylaw". Kemp told me the council believed the project was "endangering the community's lives". The council letter ordered residents to remove any trees planted on the verge.

The possibility of another way to life can be a precarious storyline. In our interview Kemp said she and McNaught were clear: they would not be removing their trees, nor would they be applying for a permit. They took this stance because "what we were doing was no different to what thousands of coastal families were doing and that was having vegetation on our verge and looking after it." While some residents followed suit, others feared the consequences of litigation and applied for a permit. Kemp and McNaught investigated the option of getting their own liability insurance but received confusing and contradictory advice. The legalities of the issue became confusing, it was hard to determine who insured, or was responsible for the public space. The precarity of their position became an obstruction in their public space text.

RE-VIEWING THE RE-USE OF COMMON LAND

There is danger in an intimately digital story of home gardens that spill out into public places: both physical and digital. Perhaps the danger lay in what was communicated rather than what was done. The lens through which others view a garden project only reveals so much. Much more goes on outside the picture frame. *Urban Food Street* was not Wall Street but like the Occupy movement it brought bodies back onto the street. Kemp and McNaught never labelled themselves activists and never aimed to be political. In fact, their aim was quite the opposite: to ensure their neigh-bours all felt welcome and included. Their aim was not to decolonise thinking, nor question understandings of land or land ownership. They never explicitly set out to create a project of reclaiming the commons or "commoning" (Gibson, 2018, 7). The *Urban Food Street* story emerged from the sharing of a small amount of produce amongst neighbours and friends. But despite those intentions its digitally visual narrative arc soared from their suburban space and collided with other times and spaces, other people and matters. For the residents of *Urban Food Street* the growing of food was the matter that mattered, it was what storied their story. But those weren't the only facts of the matter.

A garden story that signals a reclamation of the commons is a hopeful, but it is also embedded with a precarious and fragile politics. That *Urban Food Street* did not explicitly aim to reclaim the commons does not mean the social-technical imagery it produced did not signal a re-turn to the commons, and all the associated stories of enclosure and resistance. The commons, says Lauren Berlant, "is incoherent, like all powerful concepts" (2016, 397) and here too complexities and ambiguities about stories that signal a commons reclaiming emerge. In Australia a settler-colonial legacy haunts any community garden story whose plots are created within Australia's land system. This means any analysis of Australian gardens is porous rather than watertight. The narrative on the screen becomes entangled with history, theory, and messy methodology. To ignore history when we consider a story of Australian land would be to ignore the stories of genocide and dispossession that are subsumed beneath the soil. The garden is entangled in good and bad, history and future, inscription and deception, people and place, human and the more-than-human, hope and despair. It is important to notice absences in Australian garden stories. They are the places on pause, the places where we must pause for breath and wonder, what has happened here, before.

The *Urban Food Street* images forced me to re-consider what an ordi-nary, suburban space was not (Griffin, 2015, 8). There were no big global environmental organisations perched on the curb; no mass gatherings walking the streets with placards. No-one chaining themselves to trees, nor blocking traffic. This was not wall street. There was nothing to Occupy. The visual *Urban Food Street* story is a very ordinary, everyday story. It is "a state of things in which something that will perhaps matter is unfolding amid the usual activity of life" (Berlant, 2011, 5). There is potential as much as precarity in this visual narrative. Hamilton too speaks of hope in the garden. It is a space, she says, where there is "potential for relations with place to be more grounded, to be otherwise" (2019, 482). Berlant also admits that while "senses of the sense of the commons" generates a "precarious politics" (2016, 398) they may also provide potential. They may "point to what's broken in sociality" while encouraging the imagining of "a liveable provisional life" (2016, 395). The *Urban Food Street* story is as much about what it "is", as what it is "not".

In Australia, the suburban verge is a domestic, everyday space. It is not quite private, but not entirely public either. Its uses are fluid, depending on the nature of the local council within which the land resides. It is often covered in lawn, with trees planted intermittently along the streetscape. The *Urban Food Street* project was one of many stories situated on and in this verge space, but also in the digital space. It is a story about neighbourhood translated into digital community. It is also a story about wandering from the designated path. I saw this path as a place where the ecofeminist imperative of "alternative, transformative discourses" might be found (Bullis, 2015). I have wondered throughout this analysis, how should I frame the use of this re-used space? *Urban Food Street*'s communication contributes to understandings of ways people are taking a stand on issues of social and environmental importance. The *Urban Food Street* images suggest what a suburban verge could or should be used for, by comparing it with what it was or is being used for. Disrupting ordinary usage brings attention to a cause, but it can also teach about a cause (Ahmed, 2019, 210). There are complexities in these simple, digital pictures. What is there is also what is not there. For a verge to exist, a suburb exists and in Australia that means existing on invaded land.

References

Ahmed, S. (2004). *The cultural politics of emotion*. Edinburgh University Press.

Ahmed, S. (2010a). Foreword. In R. Ryan-Flood & R. Gill (Eds.), *Secrecy and silence in the research process: Feminist reflections* (pp. xvi–xxi). Routledge.

Ahmed, S. (2010b). *The promise of happiness*. Duke University Press.

Ahmed, S. (2014). *Willful subjects*. Duke University Press.

Ahmed, S. (2017). *Living a feminist life*. Duke University Press.

Ahmed, S. (2019). *What's the use?: On the uses of use*. Duke University Press.

Alaimo, S. (2016). *Exposed: Environmental politics and pleasures in posthuman times.* University of Minnesota Press.

Apunkt Schneider, F., & Friesinger, G. (2010). *Urban hacking as a practical and theoretical critique of public spaces.* transcript Verlag.

Barad, K. (2014). Diffracting diffraction: Cutting together-apart. *Parallax: Diffracted Worlds—Diffractive Readings: Onto-Epistemologies and the Critical Humanities, 20*(3), 168–187. https://doi.org/10.1080/13534645.2014.927623

Berlant, L. (2011). *Cruel optimism.* Duke University Press.

Berlant, L. (2016). The commons: Infrastructures for troubling times. *Environment and Planning D: Society and Space, 34*(3), 393–419. https://doi.org/10.1177/0263775816645989

Bullis, C. (2015). Retalking environmental discourses from a feminist perspective: The radical potential of ecofeminism. In J. G. Cantrill & C. L. Oravec (Eds.), *The symbolic earth: Discourse and our creation of the environment* (pp. 123–148). University Press of Kentucky.

Butler-Bowdon, E. (2001). Lawn and order: Aesthetics and architecture in Australian suburbia. *Studies in the History of Gardens & Designed Landscapes, 21*(2), 108–114. https://doi.org/10.1080/14601176.2001.10435240

Cixous, H. (2007). *Insister of Jacques Derrida.* Edinburgh University Press.

Cixous, H., & Calle-Gruber, H. (1997). *Rootprints: Memory and life writing.* Routledge.

Federici, S. (2019). *Re-enchanting the world: Feminism and the politics of the commons.* PM Press.

Gibson, K. (2018). *Introduction: Food as urban commons and community economics.* University of Western Australia Publishing.

Gibson-Graham, J. K. (2006). *A postcapitalist politics.* University of Minnesota Press.

Griffin, S. (2015). *The eros of everyday life: Essays on ecology, gender and society.* Open Road Media.

Hamilton, J. M. (2019). The future of housework: The similarities and differences between making kin and making babies. *Australian Feminist Studies: What Do We Want? Feminist Environmental Humanities, 34*(102), 468–489. https://doi.org/10.1080/08164649.2019.1702874

Haraway, D. J. (2016). *Staying with the trouble: Making kin in the chthulucene.* Duke University Press.

Holmes, K. (1999). Gardens. *Journal of Australian Studies, 23,* 152–162. https://doi.org/10.1080/14443059909387485

Holmes, K., Martin, S. K., & Mirmohamadi, K. (2008). *Reading the garden: The settlement of Australia.* Melbourne University Press.

Johnston, J. (2008). Counterhegemony or bourgeois piggery? food politics and the case of foodshare. In Middendorf, G., & Wright, W (Eds). *The fight over food : producers, consumers, and activists challenge the global food system.* Pennsylvania State University Press.

Johnston, J., & Cairns, K. (2012). Eating for Change. In R. Mukherjee, & S. Banet-Weiser (Eds.), *Commodity activism: Cultural resistance in neoliberal times*. New York University Press.

Knowles, C., & Sweetman, (2004). Introduction. In C. Knowles & P. Sweetman (Eds.), *Picturing the social landscape visual methods and the sociological imagination* (pp. 1–17). Routledge.

Le Guin, U. K. (2004). *The wave in the mind: Talks and essays on the writer, the reader, and the imagination* (1st ed.). Shambhala.

Mackinlay, E. (2016). The heartlines in your hand. In E. Emerald, R. E. Rinehart, & A. Garcia (Eds.), *Global south ethnographies: Minding the senses* (pp. 153–165). Sense Publishers.

Martínez Luna, S. (2019). Still images? Materiality and mobility in digital visual culture. *Third Text, 33*(1), 43–57. https://doi.org/10.1080/09528822.2018.1546484

McDonagh, B. (2013). Making and breaking property: Negotiating enclosure and common rights in sixteenth-century England. *History Workshop Journal, 76*(1), 32–56. https://doi.org/10.1093/hwj/dbs054

McDonagh, B. (2019). Disobedient objects: Material readings of enclosure protest in sixteenth-century England. *Journal of Medieval History, 45*(2), 254–275. https://doi.org/10.1080/03044181.2019.1593629

McDonagh, B., & Daniels, S. (2012). Enclosure stories: Narratives from Northamptonshire. *Cultural Geographies, 19*(1), 107–121. https://doi.org/10.1177/1474474011427361

Moreton-Robinson, A. (2015). *The white possessive: Property, power, and indigenous sovereignty*. University of Minnesota Press.

Mortimer-Sandilands, C. (2008). Landscape, memory, and forgetting: Thinking through (my mother's) body and place. In S. J. Hekman & S. Alaimo (Eds.), *Material feminisms* (pp. 265–288). Indiana University Press.

Riessman, C. (2008). *Narrative methods for the human sciences*. Sage.

Robbins, (2007). *Lawn people: How grasses, weeds, and chemicals make us who we are*. Temple University Press.

Robbins,, & Sharp, J. T. (2003). Producing and consuming chemicals: The moral economy of the American Lawn. *Economic Geography, 79*(4), 425–451. https://doi.org/10.1111/j.1944-8287.2003.tb00222.x

Rose, G. (2014). On the relation between "visual research methods" and contemporary visual culture. *The Sociological Review, 62*(1), 24–46. https://doi.org/10.1111/1467-954X.12109

Sandberg, L. A., & Foster, J. (2005). Challenging lawn and order: Environmental discourse and lawn care reform in Canada. *Environmental Politics, 14*(4), 478–494. https://doi.org/10.1080/09644010500175692

Sandilands, C. (1999). *The good-natured feminist: Ecofeminism and the quest for democracy*. University of Minnesota Press.

Soper, K. (2007). Re-thinking the good life: The citizenship dimension of consumer disaffection with consumerism. *Journal of Consumer Culture, 7*(2), 205–229. https://doi.org/10.1177/1469540507077681

Tuck, E., & Yang, K. W. (2012). Decolonization is not a metaphor. *Decolonization: Indigeneity, Education and Society, 1*(1), 1–40.

Urban Food Street. (2016a, 3 January). *Happy detouring!* [Image attached]. Facebook. https://www.facebook.com/urbanfoodstreet/photos/a.395625327284340/444088652438007/

Urban Food Street. (2016b, 10 February). *Roadways as common ground.* [Image attached]. Facebook. https://www.facebook.com/urbanfoodstreet/photos/487343611445844

Urban Food Street. (2016c, 29 June). *Crops replace lawn on the verge.* [Image attached]. Facebook. https://www.facebook.com/urbanfoodstreet/photos/534042480109290

Urban Food Street. (2016d, 29 September). *Human traffic.* [Image attached]. Facebook. https://www.facebook.com/urbanfoodstreet/photos/567795560067315

Urban Food Street. (2016e, 9 October). *Creating verge gardens.* [Image attached]. Facebook. https://www.facebook.com/urbanfoodstreet/photos/571702993009905

Urban Food Street. (2016f, 9 October). *Gardening tools: Mundane or disruptive?* [Image attached]. Facebook. https://www.facebook.com/urbanfoodstreet/photos/571714109675460

Urban Food Street. (2016g, 8 June). *Gardening tools recall another era.* [Image attached].Facebook. https://www.facebook.com/urbanfoodstreet/photos/1307231679457029

Whyte, N. (2015). Senses of place, senses of time: Landscape history from a British perspective. *Landscape Research, 40*(8), 1–14. https://doi.org/10.1080/01426397.2015.1074987

PART IV

Letter Four (Spring)

Dear K,

As you know, when I first began the groundwork for my research the earth was hard, dry, and unstable. I began digging with bare hands to scrap away the earth and my palms still have not recovered. My fingernails splintered, and blisters bubbled like balloons on my palm. I tried smashing the earth with the dull end of an axe, but even it proved no match for rubble and rock. The only rewards I received for my labours were aching arms, some damaged pride, and tides of sweat that ran down my back and forehead, stinging salt into my eyes and leaving my shirt with dark, dank patches. There seemed to be no way through this tough and difficult terrain. Despite my hopeless digging the garden continued to grow on higher ground. To the distracted observer this place probably seemed perfectly fine. On your advice, I spent many months pointing the hose beneath the leaves each day, careful not to drown the ground in too much warm water despite the drought and the harsh, humid Queensland summers we experienced. I sprinkled fertiliser generously around the base of fruit trees four times a year. Things would bloom, sprout, or go to seed, as they should. Sulphur-crested cockatoos, their wings stretched out like a wave, continued to screech overhead, calling to each other through the mist. Rainbow Lorikeets responded in kind, their daily nails-on-a-chalkboard jabber wreaking havoc among the eucalypt blossoms. After a while I started to notice things: small moments, tiny indications that something was awry. The garden was forcing me to pay attention and paying attention meant wondering why my garden wasn't growing as well as it could or should.

Small shoots emerged, but the buds became smaller with each season. Black spots peppered my tomato leaves, curling them into listless, useless appendages before fruit could set. Other plants wilted into a churlish sadness. The food that did grow tasted bland, its flavour as faded and pale as its colour. There was something the matter with my garden.

I was certain that the key to understanding what was growing on the surface required a better understanding of the secrets lying below. My simple digging method was useless and rigid. My garden was troubled, but so was the world outside the garden borders. Perhaps this was the bloom space another Kathleen, the sociologist Kathleen Stewart writes of (2010, 340). These are the private-public spheres of affect. Along with being simple representations of a gardening life, these letters are also affective stories. They are stories that create worlds we are bound to. If, as Stewart says, this bloom space is both an "allure and a threat that shows up in ordinary sensibilities of not knowing what compels" then my research now hums in this bloom space (Stewart, 2010, 340). It contains stories and thoughts about the worlds we are bound to. My chapters are the "affective projections of a constantly negotiated common interestedness" (Berlant, 2011, 226). They matter because "affect matters in a world that is always promising and threatening to amount to something" (Stewart, 2010, 340). I see now as I step back from these chapters and scan them and my garden in unison, that much of what I have been doing is simply trying to scratch some sort of rhythm back into the page.

With Love & Fury,
XXXX

CHAPTER 5

The Planthunter: Writing the World as Garden

Reid's writing has affected my own. As she fenced off a small patch of land—a concession to the realities of growing food—I found myself fencing off my own words in chunks of hundreds rather than letting them roam free and interrupting like they had in previous chapters. My hundreds, then, reflect the way this chapter comes alive with sharp edges and thick angles: my words are gathered up and contained along overlapping, cylindrical sketch lines.

My line of desire is the road less travelled, but it is also the path most cherished. I discovered the desire line when I was searching for the lost words. The tread was faded but a line forward was clear, so I walked away from where others had gone, away from concrete and capital, and towards a more desired path. Desire lines, then, are a walking that is, "worn into the landscape by countless footsteps" (Smith & Walters, 2018, 2987). A desire line is not the designated path; it is the one preferred. I walked along this path, imagining a silent solidarity with other space users.

Writing desire lines into the soil is a joyfully mundane process. But it is not glamorous wandering. It does not receive widespread accolades. Desire lines are paths that wear the language of everyday life into the soil. They plant a common ground, but also bloom with difference, staying with the trouble rather than making trouble (Haraway, 2016). Faded desires lines reveal forgotten words hidden in the understory. It will take an effort to follow in the footsteps of Cixous to understand our text crosses invisible borders (1993, 122).

R. Mickelburgh, *The Ecofeminist Storyteller*,
https://doi.org/10.1007/978-3-031-59242-3_5

A line can be many things, but it is often a means of communication. It can be a short letter or wires connecting a telegraph or telephone with another. Desire lines that curve around the garden risk being washed away in the green. Metaphors are just metaphors when there is no meaning attached and words risk becoming lost to "cult of romantic love" (Heller, 1993, 219). *When a "cluster of promises" is planted in the soil, the garden might become awash with green. But it might also lead to an attachment that is nothing more than cruelly optimistic* (Berlant, 2011, 23).

There is a Cixousian, but also distinctly Australian, sensibility to Georgina Reid's garden writing in her website *The Planthunter* (recently re-named *Wonderground*, with a slightly expanded focus) and so this chapter re-turns me to the writing of a Hélène Cixous. In this Cixousian world, all the world is a garden (2008, 150). Each month *The Planthunter* website producesd series of written stories and visual imagery, around a specific theme. It explored the entanglement of humans, plants and the natural world, with a strong focus on gardens, through written essays of between around 1000 and 1500 words and stunning, mostly photographic, images. Amongst the essays were Reid's own digital memoirs: a series of online life writing and photos detailing her thoughts, feelings, actions and, of course, her garden by the Hawksbury River, near Sydney in New South Wales. This series of digital garden diaries was a deeply personal, yet also extremely public medi-tation, and communication, on the intimacy of place, and of what it takes to get deeply acquainted with that place as a white-settler colonial woman gar-dening on a damaged, and stolen, land. When I first read over her work I saw glimpsed of a gardener grappling with this difficult positioning. In an attempt to de-tangle this entanglement, Reid's writing itself seemed to be doing the Cixousian work of "un-forgetting, un-silencing, of unearthing, of un-blinding oneself, and of un-deafening oneself" (Cixous, 1993, 83). I wondered if this un-silencing and un-blinding through writing might also be present answers about ways to communicate a new, or perhaps re-newed, environmental consciousness, one which connects writer to reader with an essential story of the land and how to live on it.

Reid's words suggested a way of communicating the link between online garden storytelling and ecological conservation. *The Planthunter* published the work of many writers and artists, but I was drawn to Reid's own memoir-style digital essays and the way were embedded in a "deep acquaintance with some place" (Plumwood, 2002, 231). These were medi-ated "stories of the self" as Poletti (2020) might say but they their digital also existence suggested some sort of relationship building, a Plumwood-style attempt at "re-animating the world, and remaking ourselves as well, so as to become multiply enriched but consequently constrained members of the ecological community" (2009, 9) through words.

If all the world is a garden (Cixous, 2008, 150), then digital diaries I studied in the online, Australian-based web publication *The Planthunter* (n.d.), signified the writing *of* one, but also the writing *about* many. In essence, they were autobiographical accounts of her gardening life, but their digital presence—their length, staggered publishing dates, and real-time observations—spoke to me as autobiographical fragments rather than a complete story. In her examination of the mediation of autobiography, scholar Anna Poletti considers autobiography as "a cultural practice that seeks a public, an audience of imagined an unknown stranger who will interact with how a specific life has coemerged with specific media forms and practices" (2020, 12). This chapter details the way I understand how Reid's digital wordwork operates as communicative groundwork through an exploration of the way plants, people and place (both on and offline) intersect, overlap, and become entangled. Reid's writing signals a commitment not only to place, but also to contends with the possibilities of relationship between people and place. Is see this as her "deep acquaintance", a way of being and writing that operates in line with that Plumwood insistence that any "deep acquaintance must consider relationships as "two-way and two-place"; that is ways "you belong to the land as much as the land belongs to you" (2002, 230). Perhaps the other two-way relationship is the one Reid has with her reader. Her writing, and my analysis of it, considers the way her body that made a physical commitment to place, while her words make that same commitment in a textual, digital space.

THE RISK AND REWARD OF A SINGLE STORY

Focusing on a single-story*teller* as a guide for innovative communication work is also risky research work. *The Planthunter* contained hundreds of digital stories, written by a wide variety of authors, but rather than a broad study of the vast array of *Planthunter* essays I chose instead to focus only on Reid's own writing. This focus on an individual digital story*teller* risks placing responsibility for environmentalism on the individual rather than the collective: a key criticism of single-narrative environmental communication (Maniates, 2001; MacGregor, 2004, 2006). Yet, there is also opportunity for this first-person narrative to generate the *sense* of collective. The digital platform connects Reid's inside and outside worlds and so her writing is a practice that reaches beyond the individual (Dobson et al.,

2018, 11). This digital, first-person, narrative is sociologically significant. A community is created when separate individuals become linked through a shared conscious. By writing *about* the nature in her life and her strong sense of concern for the more than human *in* the digital space Reid offers up an emotional story to others to connect with.

Reid's digital diaries, published between 2017 and 2019 (and which still remain on the *Wonderground* website) are more than just an account of her life and her garden. They are places where emotion and fact are deeply entangled, perhaps not unlike a garden itself. Feelings of hope and joy are juxtaposed against the very real frustrations and disappointment of living, and seeking, a deep acquaintance with the natural world. It is a world a gardener tries to control, but remains, in essence uncontrollable. There is emotionality in texts, but there is also movement. Words are feeling, but they are also objects of feeling that "move, stick and slide" (Ahmed, 2014, 14). Emotions, in words, are moving and performative. The online performativity of naming emotions is what creates "communities of feelings" (Kuntsman, 2012, 6). There are transformative sociological possibilities within this kind of narrative (Richardson, 1997, 32). Thinking about the way words name emotions, makes understanding the creation of communities clearer. Strangers who might relate to these individual stories become emotionally bound to such a story, overcoming "isolation and alienation of contemporary life" (p. 32).

As I considered her writing in depth I began to view *The Planthunter* as a website that provides a contemporary re-reading and re-writing of plants, place, and people. Reid considers, and deconstructs, the meaning of home by considering what happens when when that place called home is a damaged place, a place entangled with the more-than-human, a history of dispossession and a legacy of ecological destruction. My analysis considers a series of Reid's own personal essays—the 'River Garden Diaries'—as well as a several of her other online, memoir-style entries. I saw this work as a contemporary re-turning to the personal medium of diaries, echoing the methods of so many Australian settler-colonial women gardeners who wanted to take note of their relationship with the land they gardened on (Holmes, 2011, 10). These diary entries signalled important stories about not only gardens but Australian culture, connection to land and women's place in the world. I suspected that Reid, through her online diaries, might also be telling some tiny, digital, stories of importance.

Uncontained Writing

The garden is a container, of sorts. By writing with and about nature, and placing that writing in the digital space, Reid's writing of the garden container, becomes uncontained. Part of this un-contain-ment emerges from the writing style itself. My initial readings of Reid's work was that she uses imagination as a linguistic gardening tool. Her "tool of the mind", as Le Guin would say (2004, 207), attempts to un-contain that contained garden space and consider the ways she uses gardening practices to get deeply acquainted with place. Reid's writing is "gorgeous", that is, it is a method Le Guin describes as an attempt to descruibe an "infinite number of beautiful, surprising, powerful, audible, visible things" (1998, 39). This embrace of a romantic, poetic-prose style of writing, is juxtaposed against stunning photographic imagery. Her writing details her thoughts and emotions settling into her new home and her attempts to establish a home garden, including a food garden. She grapples with issues of boundaries and borders; of relationships between humans and non-human; and the complexities of time. But chiefly, the diary entries consider not only events occurring in her garden and the wider world, her *feelings* in response to them. The garden, and its stories, therefore, have the potential to become uncontained, breaching their own boundaries.

Un-contained writing becomes communicative method because boundaries are breached between reader and writer, human and more-than-human. Deep acquaintances, Plumwood insists, require rich narratives and narrative subjects—both human and non-human—in order to turn space into place (2002, 231). Unearthing the power of place requires revealing the "communicativity and intentionality of more-than-human others" (p. 230). This is risky writing business. It risks the charge of anthropomorphism—where human-like characteristics are attributed to the non-human (Plumwood, 2007, 32). Plumwood defends writers with anthropomorphic tendencies, insisting that the concept has become "a policeman for reductive materialism, enforcing polarised and segregated vocabularies for humans and non-humans" (2009, 127). Continuing to fear anthropomorphism, says Plumwood, confirms Cartesian assumptions that "mentalistic qualities are confined to the human and that no mentalistic terms can properly be used for the non-human". But how to do this? How to re-present experience in ways that honour the agency and creativity of the more-than-human world? That, says Plumwood, is "a task for writing" (2007, 20).

Dreams of Desire in the Garden

In July 2017 Reid published "River garden diaries: Truth, Thoreau and design dilemmas" (2017a). In this essay she compares her move to "a house on the edge of the Hawksbury River" to that of Thoreau's "Walden-like experience". This is the place of her dreams, she says; "somewhere where there are more trees than people. That's been my criteria for the last 15 years of dreaming about leaving the city behind." Her writing-as-dreaming mimics her physical act of gardening. She digs at the land around her house, clearing away "decades of dirt and weeds", while her writing un-earths a new-old history of place. An old garden begins to emerge from the foliage as she digs, "underneath the mass of green that'd been left unchecked for over 20 years". Reid unearths "gorgeous old dry stacked sandstone walls", "stairs" and "a series of low terraces at the front and rear of the house". Reid's private property un-earthing is juxtaposed with a public word-weeding.

As her digital diary entries progress Reid begins to un-silence, unearth and an un-forget the past of her new place. In October 2017 she has written and published "River garden diaries: Tree songs and a sense of place" (2017b). The essay reveals a sense of story emerging both from the ground and from her writing. It is perhaps, as Stewart writes, an affective sense of "something coming into existence" (2010, 340). Reid writes that her truth "can be found gardening the weed infested sliver of dirt sandwiched between the bush and the river". Along with digging up the ground, there is a sense of beginning to dig into her connection with her new job. Desire begins to bloom in this emerging space-place. Reid's new home is "a dream realised" but her digging is also digging up new dreams. She "dream(s) of the day I can begin planting". If, as Stewart says, affect is the "commonplace, labor-intensive process of sensing modes of living as they come into being (2010, 340) then things are emerging not only in the garden but in and through Reid's writing. Reid writes about staring out the window, through the "ancient cobwebs" and considers the three casuarinas that "reach out across the water". They not only frame her view (literally), but philosophically as well. They are, she says, a family "living from the same resources, part of the same root system". She says, "Apparently Casuarina glauca can live for hundreds of years. I fantasise that our mother tree is ancient. That she watched over this landscape before European colonisation. That she holds its secrets, whispering them only to those who listen" (2017b).

Reid juxtaposes the growth of the casuarina trees with her personal collection of potted plants, which have migrated from the city. The imported plants, she writes, initially seem out of place in this vast "wild and neglected bush landscape". She sits her plants in a "holding bay", next to the endemic native orchids, "until their stories can be woven into the fabric of place". This holding bay seems to be a metaphor for Reid's own life. Like her plants, Reid initially feels a bit out of place and seems to be pausing to considers how she will fit into this new place. She has uprooted herself from her city life and, like a transplanted tree, it will take time for her to adjust to the conditions of her new location. She admits that at this time she is not planting or designing but simply sitting and will "probably spend (her time) dreaming". There seems to be a search for a way she can connect herself with her new home and the natural world that surrounds it. Perhaps she is searching for Plumwood's "communicative virtues", to ways to be open to the non-human "as potentially an intentional and communicative being" (Plumwood, 2002, 194). There are many ways communicative virtues can be realised. The most important, Plumwood says, is listening to the other (2002, 194).

Humans, Plumwood says, must discover ways to communicate with other species. Perhaps then, this attempt at communication must also translate not only between Reid and the more-than-human species, but between Reed, her readers and the more-than-human. As she settles into her new home Reid begins to re-write nature, contending with the difficulties of seeing (or perhaps accepting) the more-than-human as a partner, collaborator or even a co-conspirator. The rich inter-species community in which she lives allows her the freedom to both imagine and contend with moral dilemmas and care requirements that might occur when living so close to the more than human.

There is an imaginative longing for communication between the human and non-human: In October 2018, Reid published "River garden diaries: And then the snake (and the rain) came" (2018e). Here she tells the story of her encounters with a diamond python. The snake settles into her "makeshift plant nursery underneath our kitchen window", sitting amongst plants and cuttings in a shelf of milk crates and timber boards. Reid names the snake Slow Mo and genders it female because "it doesn't matter, except that I feel it's kinder to say her/him instead of it". The snake arrives during a time of despair and deep desire for Reid. There has been little rain and her "soil was so dry even the weeds were struggling". She describes her makeshift garden as "150 small pots of hope and beauty", which she has stockpiled "in a wallaby-proof enclosure". This garden, she says, had been a "daily reminder of my sometimes ridiculous optimism, my romanticism". As she begins planting out the pots, the snake arrives and rain arrives soon after. In response to these natural-world visitors Reid describes herself gardening "like a mad woman". "And like the weeds, I began photosynthesising. Hope, and chlorophyll, had returned, heralded by a very slow-moving diamond python called Mo" (Reid, 2018e). These romantic imaginings allow Reid to break down affectively what is often so difficult to do practically. Her writing seems to align with the communicative virtues Plumwood ascribes for a deeper acquaintance with nature. In particular, "an openness to the non-human other as potentially an intentional and communicative being" (2002, 194). Reid's deep acquaintance is clear in this essay. She spends time just sitting, staring at the snake. Fear dissolves through her writing as she considers ways this more-than-human creature moves at its own pace, in its own time.

Just as the snake sees no borders between national park, garden, or house, Reid too blurs the boundaries between herself and the non-human world. Her sense of photosynthesis dissolves her body into a larger system. Nature is still *out there*, but it is also part of her. Photosynthesis occurs in Reid's garden, but in writing about it, Reid also breathes life into the digital space. The light she has harnessed *in* the garden as provided energy *to* garden, but also to write and to post her words in the digital realm. Her words slowly weave an entangled human, non-human, online and offline metamorphosis. She is not only being in nature; she is becoming *of* nature. She has become plant-like, rather than the snake becoming human-like. She attempts to read, and write, the world through the snake's eyes and in doing so becomes part of the ecology of place. She "sat and stared and stared"; and "learned lots of things by looking at her" (2018e). Plumwood suggests a communicative stance involves considering the "canny" animals who might "gaze back, size you up and tell you who you are" (2002, 177). Reid seems to have both discovered, and considered, the snake as her own canny animal in her home. As they have gazed at each other Reid has attempted to write a common language of stillness, slowness and attentiveness. There is a consideration of time within this narrative. A message of slowness, of loving deliberation. It is writing against the grain of short attention spans and quick judgement.

Obviously, Reid cannot communicate with the snake verbally, and so she sits quietly, observing and thinking and, perhaps, listening for a common language. The only way to avoid getting lost in translation is to spend time, *in* the garden, *with* the snake. Philosopher Michelle Boulous Walker suggests that a philosophical practice of "slow reading" can "counter the effects of containment and mastery" in philosophy (2016, 2). This too could be translated and applied to Reid's method of communication: between her and the snake, and the story she later communicates online. In taking time out to watch and read the snake she writes the snake as something like kin. The snake is not stereotyped reductively as one of the "mindless objects, non-intentional mechanisms with no potential to be communicative and narrative subjects, as lacking potential viewpoints, well-being, desires and projects of their own" (Plumwood, 2002, 175). Instead, Reid considers the snake's well-being, desires and projects of its own. The snake determines its own direction of travel and Reid does little to stop her; she is doing the opposite of stopping by gently moving some plants out of her way. Reid seems to use words as a way to both imagine and observe; they smooth the path of her relationship with the snake. They are pleasurable observations, which begin to reveal ways of a connection and communication with the more than human world.

Fear and anxiety are absent in Reid's snake story. The snake was not written as a threat, or something that was out of place. She did not try to kill the snake, nor even remove it from its place near her home. There is a sense of the "oddkin" that Haraway urges us to make—that is, finding ourselves in unexpected "collaboration(s) and combination(s)" and doing so in a way that is situated "someplace and not noplace, entangled and worldly" (2016, 4). The snake is the character, the communicative being, in Reid's garden story. Making kin with the snake is a joyful act. What most people might experience as anxiety when being situated in such proximity to a snake is replaced with wonder and joy. Reid's home is its home. Wonder replaces fear, pleasure replaces anxiety. This writing with wonder gestures towards an alternative paradigm. It brings Reid, and the reader, into relation with the snake. The reality is that the snake does not see garden borders and, through her writing, neither does the gardener. Stacey Alaimo suggests such a paradigm requires emphasizing "human dwellings as habitats, revealing our interconnections with nonhuman nature and the possibilities for a multitude of sustainable pleasures" (2016, 30). Alaimo insists we must recognise and make space for animal cultures, memories, pleasures, and homes within human landscapes, "as it is no longer possible, within the Anthropocene, to imagine they will survive somewhere else" (2016, 30). Reid writes the snake into her home and into the world. Her home becomes the place of great narrative attentiveness.

Reid tells a communicative story by weaving storylines between the appearance of both the rain and the snake, and the rebirth of Reid's garden. There is a sense of reanimation in both. Reanimation, says Plumwood, requires "hearing sound as voice, seeing movement as action, adaptation as intelligence and dialogue, coincidence and chaos as the creativity of matter" (2009, 9). There is nothing passive about the snake, but agency does not equate with aggression in this narrative. Rather, it is replaced with a sense of wonder which transports the reader to another, less familiar, world. There is romanticism emerging from her words, but not in the traditional way. Romanticism often encourages individuals to "constrain" their interactions with nature to conserve it. Such constraint, argues Chaia Heller, encourages a estrangement, rather than connection, between humans and nature (1993, 227). Reid does not constrain herself from the snake, but she doesn't intervene in its travels either. Reid on the other hand considers a life lived *with* the non-human. A media garden story such as the one Reid presents transports the reader into another narrative world. Reid tells, through her river garden narrative, of a life far from suburbia or the city. Through her digital story, readers are transported towards, what would be for many, the desired state. While it may not be the Doolittle-style "talk to the animals" stereotype, it is one that transports the reader from the everyday world to one full of desires.

River Garden Diaries: Fence Lines Around the Present

Reid's work makes me think about the way fencelines are the place in between. They slice a line between properties, indicating where property is owned, but not necessarily where it is cared for. Fencing closes off something, it separates, it encloses, it excludes and contains. But one can also stand on the fence, becoming a fence sitter with one foot on each side. Fence sitting is a feeling of uncertainty and indecision, of not knowing whose side to take and so not taking either. Fencelines mark a line in the soil between the garden and the bush. They do the work of cutting together apart (Barad, 2014).

My bones ache as I perch precariously on the old wooden palings. Sitting on the fence, being a fence-sitter, is a hard, uncomfortable place. It is the space in-between. Discomfort bares down on me as I stare longingly at the soft earth on either side. One has been cultivated into garden beds, the other composted to life with leaves and forest bark. It would be more comfortable to take a side and take up a space. But then I must consider whose side it is that I am on. What happens if I want to see both sides of the story?

Researchers are fencesitters, and fencesitters like me have a good vantage point to watch the way barbed wire cuts a property in two. At first these two halves seem mutually exclusive, contradictory. But there is potential in this position. Fences might also be "a lively affair…one that troubles dichotomies, including some of the most sedimented and stabilized/stabilizing binaries" (Barad, 2014, 168). *I have a clear view from the fence to watch seeds blowing in the wind. They settle on the unsettled land and sink into the earths as much as their heart desires. Seeds don't recognise fencelines.*

Through her narrative Reid reveals the way her garden expresses her desire: a joyful life full of inter-species entanglement. To do so she writes along lines carved out by Donna Haraway by stressing the importance of making kin with those who are not blood relatives. The trees, the plants, the animals—are very much Reid's heart's desire (Haraway, 2016, 103). Yet for all the potential and promise within Haraway's urgings, the ground along this much-desired path is uneven and difficult to traverse. There are complicated realities to consider when living amongst non-human kin. Reid's writing re-turns me to Barad's words, that "entanglements are not unities, they do not erase differences" (2014, 176). Reid might long to entangle her garden with the wider land, and for a moment she succeeded in doing so, but that has not created unity. Reid's romantic narrative attempts to dissolve the invisible border between property and national park. She initially rejoiced in the romance that a lack of fencing on her property brings and the fact the only access to her home is via the river, by boat. In March 2018 Reid published "River garden diaries: Neighbourly relations" (2018a) on *The Planthunter* website. She begins the essay full of hope and writes, "It's just the river…our little patch and the national park. No borders, no walls, no confinement. Freedom!".

Reid's fence-less place is the much-desired space. In this fence-less, joyfull story the garden is a space of deep connections to the wider area's ecology. Possibility fills her garden world, a radical alternative to the commodified dream of contained property ownership versus fenced-off special, protected lands. Protecting special areas such as national parks while commodifying, and un-protecting, areas outside those spaces is deeply problematic. Special, protected lands, says Plumwood, rely on, and are interdependent with the ordinary, commodified land. There should be no carving out of difference between the two. Creatures cannot see these divisions. When only exceptional areas can be protected other areas, which may also require great ecological protection, are forgotten. Reid writes along these lines with her fence-less garden. There is potential in this life of peace and beauty. Reid initially thinks this fence-less place is "romantic and wonderful" and her writing joyfully reflects the garden's fence-less-ness. Property boundaries that separate the special lands of national parks and with the commodified land of private property become faded and difficult to decipher. Her writing and garden space are entangled in this freedom. They are both uncontained and spill out on to the wider land and through the screen.

As Reid's March 2018 "Neighbourly Relations" essay progresses, she reveals the reality that has dimmed her initial fence-less enthusiasm. Reid is glum: "the romance has died. It's been chewed up and shat out by the horse of hungry wallabies who have discovered my penchant for planting juicy, tasty things" (2018a). She is weaving a narrative that attempts to make kin with the natural world around her. But making kin with the nonhuman is more than a surface level PR campaign. Berlant's warning that "optimism is cruel when the object/scene that ignites a sense of possibility actually makes it impossible to attain" seems pertinent here (2011, 2). Romantic imaginings might be full of hope and wonder, but like most familial entanglements, they are also complicated and often cast a veil over reality. Blurring private ownership with special lands does not equate to utopia. Making oddkin with other species is not always a method that is easily able to ignite kindness. Romantic green narratives of place can be a high-risk love story. Reid's longing for a space free of borders and containment exposes the cruel optimism of Reid's garden story. She writes: "All the cuttings I had so carefully nurtured were sticks in the ground. Sage, gone. Sedum, gone. Rosemary, gone. Cuttings were pulled out of their pots, succulent leaves were nibbled, geranium plants were stripped bare" (2018a). Heartache often follows on hope's coat tails.

Reid's writing becomes a testing ground for desire versus reality. Her immediate reaction to that "botanical destruction" of her garden is to protect it, to fence it off. Reid writes that tears flowed when the wallaby-destruction became too much. Reid writes, "all of a sudden, I was building a fence in my head. A big fence, enclosing the entire property" (2018a). Many gardeners would understand Reid's feelings; they would feel the devastation as keenly as if it had happened to them. (It probably has.) A gardener feels something (usually anger, often disappointment) when carefully tending seedlings and cuttings over days and months into bloom only to have their hard work devastated by a more-than-human critter or weather event. It takes care and time to nurture a seed to flower. Destruction is also loss. I have, like Reid, shed tears after discovering a once abundant row of parsley or tomatoes nibbled to nothingness by possums in the night. In the dreams of gardeners, desire often outweighs the need to endure. Yet enclosing the garden from the wider, non-human world follows the settler-colonial path and, despite her heartache, she seems hesitant to follow that route. Fencing installation poses a risk to romantic garden narratives.

Reid's writing sits stuck in between all the pieces of the fence that are yet to be built. In August 2018 Reid published "River garden diaries: Composting jellyfish and other meanderings" (2018c) on *The Planthunter* website. She writes of finding "a big roll of fencing wire for $10 at a junk shop and decided it was a sign that I needed to fence our entire property. All 1500 square meters". The botanical destruction has forced her to ask herself hard questions about herself and her position, as a human, in a more than human world. The garden becomes a precarious place that is part of a more than human space. Answers do not emerge from the soil easily. She writes in this essay:

> I surprised myself. Here I was, a woman with a head full of romantic notions of creating a garden that connects so beautifully with the native bushland that it would be hard to tell where it began and the bush ended; yet suddenly I was planning to build walls to keep the wilderness out. What? Why? (Reid, 2018c)

Despite the gardening setbacks, Reid's essay wanders with possibility rather than despair. She considers building "walls to keep out wilderness", but concedes, "a garden is not an island. Imposing oneself on the land does not a good garden, or gardener, make" (2018c). Reid wonders how it is possible to relate to, and care with, the non-human other in a way that suits all of them. Her desire to grow plants, to control where and when they grow, prompts her to consider the divide between public and private property, or between what is considered special (national parks) and not special, or unprotected and so commodified land. Plumwood might say, "the precarious balance of sameness and difference, of self and other involved in experiencing sameness without obliterating difference" (2002, 199). Reid would still *like* "a wild and abundant garden" which provides food for her family as well as the birds, bees, insects and even the wallabies. The object of Reid's desire—the entanglement of her garden and the wider property—remains. However, it is obvious that this cannot be achieved without compromise.

A deep acquaintance with place and an inter-species life has exposed harsh realities that romance often silences in a gardening storyline. Rather than fence off her entire property (and possibly at the same time her heart's desire) Reid's writes in the August 2018 "Composting jellyfish" essay that her parents convince her to narrow the scope of her desire. She decides to use part of the fencing for "a small, yet secure, patch". It is a small concession, yet this fenced patch of land becomes, "an island of potential among the desolation of the rest of the garden (though garden is, perhaps, too strong a word for the land under my care)" (2018c). She refuses to surrender the romance of this place, or in her writing. The fence is there but, for now, it seems temporary. The wider garden remains uncontained. She insists on exploring "notions of both romance and pragmaticism and somehow find a pathway between the two". In doing so, hope returns to both her heart and garden. The fencing might have forced her to backtrack on her initial reluctance to reduce her garden to an island. But, for now, it seems to provide a physical, but also linguistic, vantage point for surveying the scene and getting more deeply acquainted with place.

Reid might have solved one problem through enclosure—protecting her seedlings from hungry, non-human species—but she also faces the other problem of attempting to grow food in hard, nutrient deficient soil. Like most vegetable gardeners, Reid must consider the health of the soil on which she wants food to grow. The soil might be fine for growing endemic plants, but not so fine for growing edible plants. To rectify the problem, she adds "bag upon bag of cow poo, layer upon layer of lucerne mulch, and can't make compost fast enough" (2018c). She considers everything as "fodder for the compost heap" including the dead jellyfish that wash up at low tide. Her composting is transformation; she has evolved "from modest compost appreciator to decompositional fanatic" (2018c). This story of the compost weaves a narrative where words become entangled in the decay of living and nonliving. Death works to create life. Her essay travels in a circle again, once more considering the hopefulness of her position, as well as the disappointment. She writes:

> We've been promising chickens to my partner's kids since we moved here. Recently I've been suggesting quails as an alternative because they're smaller and shorter lived than chickens, therefore easier to bury in a compost heap when their day comes. My step-kids were suitably unimpressed by my rea-

soning. They think I'm dreadfully morbid. All I can see is beautiful, dark, life sustaining compost. (Reid, 2018c)

A simple narrative about building compost and soil is, like the composting process itself, working harder than it seems. It reconceptualises understandings of non-human agency, sensitising the reader life in a multi-species community. Reid's words about dirt demonstrate an "agency without agents, a foundational, perpetual becoming that happens without will or intention or delineation. In fact, dirt—a rather indiscrete substance—is necessary for the emergence of less diffuse life forms" (Alaimo, 2008, 247). Dirt, in Reid's narrative, becomes an "unlikely candidate for glory" (Alaimo, 2008, 249). Ladelle McWhorter points out, soil acts and aggregates, arranging itself in empty spaces, creating water and filtration systems, creating more dirt, and supporting life that turns dead matter into dirt (1999, 167). Dirt "perpetuates itself" and "life never surpasses dirt, because life rides on dirt's coat tails" (p. 167). Death sustains life, but compost is not passive. It is a participant and respondent to change, created by humans (Merchant, 2010, 8). Reid's narrative illustrates how dirt breaks down death, but in doing so contributes to further growth and life.

LOVELINES TO THE FUTURE

I began avoiding my own garden while I studied these words of another women. Not enough time, water, or love. This busy forgetting, at first, seemed practical. There's no need to feel guilty when you're busy forgetting. But the garden became so overgrown, elephant grass scraped my knees as I walked through and small animals sheltered in the undergrowth. Deborah Bird Rose whispered, "if we are to forget all the things that make us uneasy, then we have to forget even to think about the places where uncomfortable things happen" (2004, 47). Gardens can become an uncomfortable place when words lose their meaning.

Forgetting means losing something you've neglected. Forgetting is also "insulating ourselves against the absences that surround us." I imagined her surveying my gardening, narrowing her eyes at the way I had busily forgotten "what is vulnerable, endangered, under threat" (Bird Rose, 2004, 47). To search for these lost words then is to imagine an alternative but it is also digging up ways un-forget. The media has all it needs to create a community of the imagination, they're just not doing it (Le Guin, 2004, 209), and neither was I.

Unsettling doesn't need to busy herself forgetting. She flies further than any fence. Feathers float in her wake, drifting to the ground like a trail to follow. I pick up these feathers and think about the love they bring as I walk along the path. Words of love might erase physical fencelines if you imagine them carefully enough. Unsettling has shared her sentiments, but it is up to me to decipher the code she speaks in. A participatory culture can be a communion of minds when words are unearthed together. A garden, then, becomes a place of mutual expression.

Reid's romantic imaginings, immersed in inter-species reality, gestures towards a peaceful, loving alternative to the capitalist demands of everyday life. She loves her new home: the river that flows alongside her land; her garden; and the national park. This love emerges slowly, as one might weave a basket. Each line becomes interwoven with the last and will be connected to the next. It resists the need to rush and the everyday busyness that results in overlooking the interconnectedness between human and the natural world. In July 2018 Reid published "River garden diaries: A Book in a Boatshed and Pondering Place" (2018b). In this essay she writes that "as I've written, I've fallen in love with this place". Reid strives to develop a language to adequately describe her embeddedness in place. She writes about having "grown roots in this patch of dirt next to the river"; getting to know "the rhythms of the bushland"; and feeling more grounded. Her narrative moves from an individualist story of *me and them* to an attempt at a more enmeshed narrative of *we*. A line flows through the *we* and joins with other words and metaphors of love, care and responsibility. Her writing draws a love line to the land. It is a multi-layered, complicated story of home love. She highlights the importance of loving place as a act that is full of care, writing:

> Maybe just being in a place, loving a place, and acting on behalf of love for that place is a more powerful idea than I once thought? Maybe it's the best way to express love and care for the world—a love grounded in both truth and lived experience? (Reid, 2018b)

Reid writes of her garden with words of love, but like most things to do with gardening and relationships, it is a complicated devotion. The narrative line of love that connects Reid's garden with the wider landscape allows her to cross borders of the place she lives and works. There is love, but there is also despair. On 17 October 2018, Reid published "The way of the gardener" (2018d). Referring to the recently released report from the UN Intergovernmental Panel on Climate Change, Reid notes the consequences of temperatures increasing by four degrees "if nothing changes". She highlights the dire predictions: the extinction of the Great Barrier Reef; the loss of plant, insect and invertebrate life, food and water insecurity and extreme weather. Her garden is one small story in the much bigger, many storied world of gardens and nature and a human's relationship with nature. She is, perhaps, speaking of the "small stories" that Melissa Lucashenko says "might help us find a way through" (2003). Through her small garden stories, Reid observes, and expresses being "directly engaged with the health of the ground beneath my feet". she rejects linear time, linear growth. Instead, in the essay she writes,

> I know from experience that decay enables life, and that a healthy garden requires attention, faith, action and care. Each day I am reminded to respect the laws and the whims of Mother Nature. I cannot ignore the beauty and wonder and madness and pain of the world because I am engaged in its care. I live in relation to it. I love it, and today I grieve it. (Reid, 2018d)

If Reid's garden is a story of love, then that love requires facing uncomfortable facts. Her garden story forces her writing to drift from her own patch of land to the larger issues at hand when gardening in Australia. It is acknowledging that her gardening happens at a time of climate and ecological catastrophe, and within a legacy of dispossession of, and violence towards, Indigenous people. She is questioning the usefulness of the glocal. Reid is a white, settler colonial woman curating a garden on private property. As she grapples with ways to build connectedness between people and place, she contends with issues of language and loss. Her essays begin to grapple with her complicated positioning on the land she loves and gardens. In "Say my name: On speaking the Indigenous names of plants" (Reid, 2019a) Reid writes about an emerging desire to learn, and speak, the names First Nations people of the Sydney area gave those same plants. She acknowledges that "within these words is a world, a way of seeing vastly different to the one I've inherited. A way that has grown from the land itself" (2019a). She quotes Professor Jakelin Troy, a Ngarigu woman and director of the Aboriginal and Torres Strait Islander Research at the University of Sydney, who says naming is "not simply about classification but acknowledging connection". Reid wonders what happens when country is known, not simply from a place of classification and objectivity but also a place of connectedness. She wants to begin using and remembering Indigenous plant names because "they're words that reflect the kind of relationship I want to cultivate with the land I live upon, and they're words that respect the deep connection the First Nations people of Australia have with the country, through language" (2019a).

Reid's garden story is incomplete and unfinished. There are loose ends that have not been, or perhaps cannot be, tied up neatly in a bow. Reid might be entangling herself with her area's history, but unification still seems to elude her. She admits her knowledge is incomplete; her ignorance around language and place "has the potential to diminish the complexity and nuance of the relationship in the first place." It is a concern that echoes much of Bird Rose's writing as well. While entanglement can be "grounds for action" and dialogue as a space for "love, respect, sympathy and a determination to act together" (2004, 22) for now Reid's entanglement still lacks unity. Yet remembering and speaking Indigenous plant considers the silence that has smothered Australian soil, that "big swag" Bird Rose says lies between Indigenous and settler-colonial Australians (2004, 22). Reid is not just cultivating plants, but a desire. She seeks a renewed relationship between present time and a previous time, when others cared for the land she is now on.

The word 'garden' allows Reid to imagine and write an alternative paradigm, to communicate global environmental and social issues from a local perspective: the place of her garden. Her re-turn to the word garden as another word for her love of the more-than-human. In her August 2019 article "Audacious gardening: On daring to care" (2019b) Reid reveals that she initially avoided the word gardening in the messaging of *The Planthunter* because she "wanted to be taken seriously", and, more specifically she was "more interested in the why than the how". Writing several years since the launch of her publication, she reclaims the word "garden" as a way to describe caring "deeply, inclusively, and audaciously for the world outside our homes and our heads." For Reid, the word "garden" is a promise, not a threat. Rather than describing what gardening does, she describes what it is. That is, it is "to care deeply". The is a sense of the reweaving in her narrative, a weaving with "beauty, connection, growth". Her commitment to the word "garden" grows, "with every news report of impending climate catastrophe, every new plant placed on an ever-expanding endangered species list, every time a politician justifies a project with potentially disastrous environmental outcomes as 'good for the economy" (2019b). Reid expresses her deep and loving acquaintance with some place by reinstating the word "garden". For her the word "garden" is a framework "grounded in care and action" to engage with the world.

Taken together Reid's writing both searches and suggests what might constitute a deep acquaintance with the garden place she writes about. The digital diaries are a quest for deeper connection to both the social and the spiritual. It is a search to understand land as relationships that are "two-way and two-place, in which you belong to the land as much as the land belongs to you" (Plumwood, 2002, 230). This dual gleaning is perhaps a pursuit for mutuality in the garden. If mutuality is the "generous sharing of resources" to express love then the garden becomes that place of "giving and receiving" as an everyday ritual (hooks, 2001, 164). Reid too considers the presence of mutuality in the place of her "ritual of habitation" (Holmes, 1999, 153), the garden space.

Reid's loveline, then, is also a digital love letter to the world. It communicates her love for her garden world, but also a love for the wider world through words about her relationship to the land and its more than human inhabitants. There is a sense of advocacy, of persuasion and connection with others, within this individual narrative. Ahmed warns against assuming that love is all that is needed for a foundation of political action or for good politics. However, she concedes that while love does not need to shape visionary politics, we can however love the visions we have (Ahmed, 2014, 141). In understanding this "we can find perhaps a different kind of line or connection between the others we care for, and the world to which we want to give shape" (Ahmed, 2014, 141). There is a sense of "affective solidarity" that takes shape within such lines (p. 141). Considered together, her memoir-style writing recognises that there are conditions and troubles associated with this love. Her loveline is entangled with fence lines; whether she likes it or not the physical boundaries and borders of her garden remain.

Through writing of the place where fencelines and lovelines become entangled, Reid also textually imagines that somewhere within that entanglement there is a garden gate. Her writing seems to open that gate to global readers and encourage them to wander into, but also past, the containment of her garden. Her garden, her object of desire, is entangled with pleasure and wonder throughout poetic-prose. Pleasure emerges through emotional *and* sensory stimulation. It is a pleasure to read—a romance-reality story between woman and her garden. Reciprocity blooms in her writing. Wall Kimmerer suggests the knowledge of loving the earth changes you, but "when you feel that the earth loves you in return, that feeling transforms the relationship from a one-way street into a sacred bond" (2013, 125). To read and listen to these garden stories is to sit for a while with hope, but also with reality. There is enjoyment in this mediatised experience. The reader's imagination is freed from the constraints of everyday living, social pressures, and the guilt of increased consumption. Love might grow old in a garden, but it can also be transplanted anew onto the line on the screen.

Gardening in a Shared Habitat of Media Ecology

As I read Reid's work I re-turned weekly to the compost where scraps of something and nothing lay waiting. I twisted and turned the mess and my imagination, encouraging it to decompose into another matter. Family dinners, lawn cuttings, chicken poop, leaves and even shredded paper from old drafts swirled together and each time I re-turned more worms emerged. These scraps of the human and non-human were objects of both waste and value and soon my words about gardens dissolved into the garden I grew. Their disappearance did not mean there was nothing there. It just meant they were part of something bigger.

Once I began using that imagination tool to aerate my compost my garden grew like an old growth forest. My desire was entangled in fencelines but I did not go back into the past but rather turned it over again and again, breathing new life, troubling dichotomies as "a singular act of absolute differentiation, fracturing this from that, now from then" (Barad, 2014, 168).

My imagination also issued a warning: metaphors are porous and poetry is leaky. The garden is the place in the space, the idea, and the action (Francis & Hester, 1990), *but it is also the digital place. It aged well in secret, but the real secret was that it could not be contained with borders and fencing and hedges. A metaphor can never just be a metaphor. It must also do something. My hunt for these lost words of feelings and affect are in the dirt but have also been transplanted on to the screen. Combined they create the culture my words are embedded in.*

Considering what matter matters (Haraway, 2016, 12) is to consider the temporal strangeness that emerge from Reid's essays. Georgina Reid's words emerge from the physical, natural space of the garden soil. But they are also transplanted into the opposite of that tangible place: the technical, immaterial, online place. By placing the stories of her private garden into the online world, Reid's small, private garden reaches far from the shores of her river home to a global audience. Reid writes about a private place but when she places her words about that place in the digital space, it becomes a public story that affects her readers. Her words about gardens become an affective space for readers around the world. But her stories are not just about the physical act of gardening, the "ritual of habitation" (Holmes, 1999, 153). Themes of desire, boundaries, and love weave lines through Reid's circular storytelling: her storylines. They build as time goes on, but they also circle around, double up and contest and critique each other.

While they might be written over time, Reid has planted her words in the online world where they remain, for now. Doing so lets the reader access any one of them, at any time, in any order. Reading the essays over time, but not necessarily in time, disturbs her personal and ecological narrative time-line. Of course, they could also be removed at any time. In "River garden diaries: A Book in a Boatshed and Pondering Place" (2018b) Reid describes the building of her garden as a "journey of affection" that requires both listening and seeing. The digital dynamic of Reid's narrative journey is caught in, but also seems to be attempting to disentangle itself from, Bird Rose's "web of time concepts" (2004, 18). The digital dynamic of Reid's writing allows her to attempt what Bird Rose might call "recuperative work"; that is, writing where the linear past, present and future is broken down and instead imagined as "all accessible time…rich with possibility" (p. 25). Alternatives to linear time include considering "the time of the generations of living things, including ecological time, synchronicities, intervals, patterns, and rhythms" (p. 25). These temporal possibilities enable the reader to dream of another world, another way to be. Reid's words placed in a networked world makes her garden not only one made of soil but also a series of connections with strangers that continue long after her stories are published.

Reid's work illustrates how online garden stories are a socio-technical relationship with nature, where matter is the soil, but what matters is also the digital space. Donna Haraway's words ring true. That, "it matters what matters we use to think other matters with; it matters what stories we tell to tell other stories with; it matters what knots know knots, what thoughts think thoughts, what descriptions describe descriptions, what ties tie ties" (2016, 12). Reid's garden is also a garden metaphor about wider ecological and social justice issues. The platform is the medium, but the medium is also the soil and the words hidden within. In gardens the soil is the culture; it is a prepared medium. But culture can also be about being embedded in something. Reid's essays both appear in an online culture and emerge from a soil culture. They exist within what Jodi Dean calls the "technoculture of democracy", that is amongst lives lived in the information age, governed by publicity (2018). Such technoculture "relies on the conviction that the solution to any problem is publicity" (p. 15). In The Planthunter too there is a sense that the writing wants to *do* something. The language gestures towards an alternative life to contemporary capitalist demands that are so detrimental to both the human and non-human environment. There are underlying arguments within the narrative: the importance of human-nature connection; the rapid degradation of the natural world; how we might, as individuals, make change. Reid relies on a networked public for these stories to reach others. She is not just writing for herself.

Reid's written words embrace the joyfulness of an alternative hedonism, embedded within an online, communicative capitalism. There is deep complexity within this placement. Words in and around gardens remain embedded in the very same capitalist processes that have inflicted so much harm on the environment. Embedding garden words in a technoculture risks burying publicity's capitalist secret in the soil. They become a secret world of words that are both powerful and not powerful enough. Dean points out that the very nature of the web can stifle political action. Within its communicative, capitalist, networked space trust becomes "displaced from persons to technologies", where the "pressure to know, to find out for oneself, to be informed sucks the life out of political action" (Dean, 2018, 164). This postponement of action, says Dean, "is a permanent deferral" (2018, 163). Why bother writing, and perhaps also analysing, yet another narrative that, by the medium through which it is produced, creates no substantial changes or improvements to the dire ecological situation humans find themselves in? There is a glimmer of optimism in the technical narrative. Dean admits that although the web is a "site of conflict" it is also a "source of democratic potential" (2018, 166). This potential, suggests Dean, occurs by seeing the Web as not a tool or communication medium but rather a series of "issue networks" where "flows of communication and contestation that turn matters into issues" (Dean, 2018, 169). In other words, garden stories affect how readers think and feel in the world. That in turn shapes what they do in the world.

Georgina Reid's voice is clear, but the line we talk on for our interview is scratchy. I cannot see her, but the hours I've spent poring over her memoir-style essays about life in a small cottage by the banks of the Hawksbury River allow me to imagine her in place as we chat. My response to her writing has unfolded like a mirror; it straddles the fence in the same way her words do on the screen. She writes of a new life and garden and home on the banks of a river but untangling the meaning within has been difficult and often at times contradictory. As we make small talk before the formal part of our phone interview commences, I hear Reid's dog bark in the background. Having immersed myself in her stories of a garden by the river, I imagine small waves lapping on the river shore. Phone contact been the only way we have been able to have our conversation. Although most of my analysis here focuses on her written, public work I felt the need to talk with her to fill the spaces where unanswered questions lay. Specifically, how did she try to resist the very system that she needed to produce her publication. At the time of our conversation Reid revealed she had a difficult relationship with advertising revenue. She had experimented with "a bit of advertising", but by the time we spoke she had decided to stop all paid advertising. She felt it didn't fit with her vision for The Planthunter. Her writing style, she said, didn't lend itself to commercial advertising. It was a struggle then to commercialise *The Planthunter* "in a way that felt right". But she still needed some way to continue its operation. Her solution was two-fold: one off or monthly donations as well as a business directory. Businesses could "show support" through The Planthunter directory, where "likeminded brands and organisations" could connect with an audience. Many of those who participated were relatively small, niche businesses specialising in horticulture, landscaping, and growing or producers of creative, art-based products.

The deep complexities and ethical dilemmas Reid faced to produce an online narrative are not unusual. Its situatedness in the online space means that *The Planthunter* will always be entangled in the complexities and ambiguities of the networked, platformed and therefore capitalist world. The Planthunter reflects the way creativity is inherent in contemporary eco-culture, where citizenship and connection often rely on not only romanticism and nostalgia, but also the capitalist, networked world. Reid's story parallels these competing, and often conflicting, demands. She aligns her vision with the romantic focus on "aesthetics, pleasure and the art of everyday living" (Lewis, 2012, 323). They signal a shift from a green political culture out there, to "its diffusion into every aspect of people's daily lives" (Lewis, 2012, 316). Green lifestyle choices in domestic spaces reflect the individualisation of global environmental concerns. Green citizenship creates a link with "creativity, community-building and romantic concerns about the art and aesthetics of everyday living" (p. 316). These shifts, "speak to broader trends in late modern suburban nations like Australia where a range of forms of environmentally oriented consumer and lifestyle based "activism"—from community gardening to food co-ops—are currently reshaping the nature and meaning of citizenship" (p. 316). Reid's self-confessed romantic imaginings she represents a style of what Kate Soper calls "the other pleasures of post-consumerism" (2007). Pleasure, wonder and joy, hold significant, untapped potential within environmentalism.

While there is much to celebrate within The Planthunter, I find myself wanting to proceed with caution. Romance can hurt and the desire to protect often requires an imagination filled with weakness and vulnerability. I re-turn to the path that carries Berlant's cautionary tale about optimism, that it can be cruel; an object of desire is also a "cluster of promises we want someone or something to make to us and make possible to use" (2010, 93). The problem of "cruel optimism" occurs, says Berlant, when there is "a relation of attachment to compromised conditions of possibility whose realization is discovered either to be impossible, sheer fantasy or too possible, and toxic" (2010, 94). It is difficult to see how a story like Reid's, which considers the natural world and a human's place in that world through the lens of a home garden, fits into this cruelly optimistic narrative. Of course, the risk is that if all the reader is doing is reading and imagining and dreaming that something might happen, then nothing at all will happen. Placing the responsibility of environmental protection on individual behaviour rather than global, structural interventions might simply result in an optimism that is not only cruel but fails to truly allow environmental protection to flourish. The romanticism embedded in these stories could mean they represent nothing more than a dream. Yet the digital dynamic, entangled with a language of love in Reid's text holds the promise of collective emotions, and in doing so, the potential for community. Her "some place", a home garden space, might be the spark of a story that lights others up.

References

Ahmed, S. (2014). *The cultural politics of emotion.* (2nd edition) Edinburgh University Press.

Alaimo, S. (2008). Trans-corporeal feminisms and the ethical space of nature. In S. Alaimo & S. J. Hekman (Eds.), *Material feminisms* (pp. 237–264). Indiana University Press.

Alaimo, S. (2016). *Exposed: Environmental politics and pleasures in posthuman times.* University of Minnesota Press.

Barad, K. (2014). Diffracting diffraction: Cutting together-apart. *Parallax: Diffracted Worlds—Diffractive Readings: Onto-Epistemologies and the Critical Humanities, 20*(3), 168–187. https://doi.org/10.1080/1353464 5.2014.927623

Berlant, L. (2010). Cruel optimism. In M. Gregg & G. J. Seigworth (Eds.), *The affect theory reader* (pp. 93–117). Duke University Press.

Berlant, L. (2011). *Cruel optimism.* Duke University Press.

Bird Rose, D. B. (2004). *Reports from a wild country: Ethics for decolonisation.* University of New South Wales Press.

Cixous, H. (1993). *Three steps on the ladder of writing.* Columbia University Press.

Cixous, H. (2008). *White ink: Interviews on sex, text and politics.* Acumen.

Dean, J. (2018). *Publicity's secret: How technoculture capitalizes on democracy.* Cornell University Press.

Dobson, A. S., Carah, N., & Robards, B. (2018). Digital intimate publics and social media: Towards theorising public lives on private platforms. In A. S. Dobson, B. Robards, & N. Carah (Eds.), *Digital intimate publics and social media* (pp. 3–27). Palgrave Macmillan.

Francis, M., & Hester, R. (1990). *The meaning of gardens: Idea, place and action.* MIT Press.

Haraway, D. J. (2016). *Staying with the trouble: Making kin in the chthulucene.* Duke University Press.

Heller, C. (1993). For the love of nature: Ecology and the cult of the romantic. In G. Gaard (Ed.), *Ecofeminism: Women, animals, nature* (pp. 219–242). Temple University Press.

Holmes, K. (1999). Gardens. *Journal of Australian Studies, 23,* 152–162. https://doi.org/10.1080/14443059909387485

Holmes, K. (2011). *Between the leaves: Stories of Australian women, writing and gardens.* UWA Publishing.

hooks, b. (2001). *All about love.* Harper Collins.

Kuntsman, A. (2012). Introduction: Affective fabrics of digital cultures. In A. Karatzogianni & A. Kuntsman (Eds.), *Digital cultures and the politics of emotion* (pp. 1–17). Palgrave Macmillan.

Le Guin, U. K. (1998). *Steering the craft: Exercises and discussions on story writing for the lone navigator or the mutinous crew.* Eighth Mountain Press.

Le Guin, U. K. (2004). *The wave in the mind: Talks and essays on the writer, the reader, and the imagination* (1st ed.). Shambhala.

Lewis, T. (2012). "There grows the neighbourhood": Green citizenship, creativity and life politics on eco-TV. *International Journal of Cultural Studies,* 15(3), 315–326.

Lucashenko, M. (2003). Not quite white in the head. *Griffith Review, 2,* 7–15.

MacGregor, S. (2004). From care to citizenship: Calling ecofeminism back to politics. *Ethics & the Environment, 9*(1), 56–84. https://doi.org/10.2979/ETE.2004.9.1.56

MacGregor, S. (2006). *Beyond mothering earth: Ecological citizenship and the politics of care.* UBC Press.

Maniates, M. F. (2001). *Individualization: Plant a Tree, Buy a Bike, Save the World? Global Environmental Politics, 1*(3), 31–52. https://doi.org/10.1162/152638001316881395

McWhorter, L. (1999). *Bodies and pleasures: Foucault and the politics of sexual normalization.* Indiana University Press.

Merchant, C. (2010). *Ecological revolutions: Nature, gender, and science in New England* (2nd ed.). The University of North Carolina Press.

Planthunter. (n.d.). *The Planthunter.* https://theplanthunter.com.au/

Plumwood, V. (2002). *Environmental culture: The ecological crisis of reason.* Routledge.

Plumwood, V. (2007). Journey to the heart of stone. *Nature, Culture and Literature, 5,* 17.

Plumwood, V. (2009). Nature in the active voice. *Australian Humanities Review, 46,* 113–129. https://doi.org/10.22459/AHR.46.2009.10

Poletti, A. (2020). *Stories of the self: Life writing after the book.* New York University Press.

Reid, G. (2017a, July 4). River garden diaries: Truth, Thoreau and design dilemmas. *The Planthunter.* https://theplanthunter.com.au/gardens/river-garden-diaries-truth-thoreau-design/

Reid, G. (2017b, October 17). River garden diaries: Tree songs and a sense of place. *The Planthunter.* https://theplanthunter.com.au/gardens/river-garden-diaries-tree-songs-sense-place/

Reid, G. (2018a, March 20). River garden diaries: Neighbourly relations. *The Planthunter.* https://theplanthunter.com.au/gardens/river-garden-diaries-neighbourly-relations/

Reid, G. (2018b, July 18). River garden diaries: A book in a boatshed and pondering place. *The Planthunter.* https://theplanthunter.com.au/culture/river-diaries-book-boatshed-pondering-place/

Reid, G. (2018c, August 21). River garden diaries: Composting jellyfish and other meanderings. *The Planthunter.* https://theplanthunter.com.au/gardens/river-garden-diaries-composting-jellyfish/

Reid, G. (2018d, October 17). The way of the gardener. *The Planthunter.* https://theplanthunter.com.au/culture/the-way-of-the-gardener/

Reid, G. (2018e, October 24). River garden diaries: And then the snake (and the rain) came. *The Planthunter.* https://theplanthunter.com.au/gardens/river-garden-diaries-snake-rain-came/

Reid, G. (2019a, June 11). Say my name: On speaking the Indigenous names of plants. *The Planthunter.* https://theplanthunter.com.au/culture/say-name-speaking-indigenous-names-plants/

Reid, G. (2019b, August 7). Audacious gardening: On daring to care. *The Planthunter.* https://theplanthunter.com.au/gardens/audacious-gardening-daring-care/

Richardson, L. (1997). *Fields of play: Constructing an academic life.* Rutgers University Press.

Smith, N., & Walters, P. (2018). Desire lines and defensive architecture in modern urban environments. *Urban Studies, 55*(13), 2980–2995. https://doi.org/10.1177/0042098017732690

Soper, K. (2007). The other pleasures of post-consumerism. *Soundings: A Journal of Politics and Culture, 35,* 31–40. https://doi.org/10.3898/136266207820465930

Stewart, K. (2010). Afterword: Worlding refrains. In M. Gregg & G. J. Seigworth (Eds.), *The affect theory reader* (pp. 239–253). Duke University Press.

Walker, M. B. (2016). *Slow philosophy: Reading against the Institution.* Bloomsbury Academic.

Wall Kimmerer, R. (2013). *Braiding sweetgrass.* Penguin Books.

CHAPTER 6

A 'Not' Conclusion

Gardens are rarely finished. A gardener always has more plans, questions, frustrations, hopes and dreams. Often the same can be said for stories and communication. This final chapter is brief because it is also a "not"-conclusion. Such conclusions are *not*-uncommon amongst feminists and writers, just as they are *not*-uncommon amongst gardeners. Both Virginia Woolf (1929/1977) and Elizabeth Mackinlay (2016), have written about failing to conclude. I follow their lead along this critical-creative investigation into gardening communication and, in this chapter, write a conclusion that only offers an opinion on what a garden does to help women storytellers into being and becoming more sensitive of the world around them.

I was lost for words when I began this book and I remain troubled by questions I have unearthed in my garden wanderings. Questions like:

If the garden makes us feel something then what is this feeling and what is its function?
How do I prove that it makes us feel this way?
Do these small stories say anything about the bigger picture?
What can they do about the climate emergency?
Do they do anything, or nothing at all?
Is a garden only a metaphor and is writing only writing, or is there a more complex entanglement at play?

R. Mickelburgh, *The Ecofeminist Storyteller*, https://doi.org/10.1007/978-3-031-59242-3_6

133

With these questions, there is more work to be done. Things keep catching my eye. I imagine that this is what Virginia Woolf was writing about when she wrote of "The moment". My concluding moment is composed of other moments and will compose future moments. Those sinking legs of the chair, sinking through the rich garden earth will emerge again one day (Woolf, 1952, 9). This book is but a small moment within the bigger, tougher moment of a climate emergency and the feminist struggle. It is the glocal moment, a tough moment in the tough life that writer Jeanette Winterson speaks of when she speaks of the importance of words. A tough life, she says "needs a tough language and that's what poetry is, that's what literature offers ... a language powerful enough to say how it is. It isn't a healing place, it's a finding place" (2012, 40). I add to that senteniment: that is what communication must be. This chapter *not*-concludes the active, tough work of research-writing digital garden work. I don't want to wrap this book up in a neat bow because there are still unsettling things that aren't settled. I want readers to remain unsettled, and continue to do the work needed to improve communication practices that help to protect and preserve the environment.

> To the climate change activists, this is what I say:
> Walk, as I did with Barad's refrain on repeat, that "entanglements are not unities" (2014, p. 176), but consider them alongisde Bird Rose's sentiments, that "entanglements give us grounds for action" (2004, p. 22).
> To the ecofeminists, this is what I say:
> The contemporary online women's garden stories I studied were something new, but they scored over old repetitions. In doing so they were refrains, and refrains "are a worlding" (Stewart, 2010, p. 339).
> To the academics, this is what I say:
> Something else is happening in this garden story moment. I *sense* that something that could be happening is the communication of common-sense abilities. Walk with me.

References

Barad, K. (2014). Diffracting diffraction: Cutting together-apart. *Parallax: Diffracted Worlds—Diffractive Readings: Onto-Epistemologies and the Critical Humanities, 20*(3), 168–187. https://doi.org/10.1080/13534645.2014.927623

Bird Rose, D. B. (2004). *Reports from a wild country: Ethics for decolonisation*. University of New South Wales Press.

Mackinlay, E. (2016). *Teaching and learning like a feminist storying our experiences in higher education*. Sense Publishers.

Stewart, K. (2010). Afterword: Worlding refrains. In M. Gregg & G. J. Seigworth (Eds.), *The affect theory reader* (pp. 239–253). Duke University Press.

Winterson, J. (2012). *Why be happy when you could be normal?* Grove.

Woolf, V. (1929/1977). *A room of one's own*. Granada.

Woolf, V. (1952). The moment: Summer's night. In V. Woolf (Ed.), *The moment and other essays* (Uniform ed., pp. 9–13). Hogarth Press.

References

Ahmed, S. (2004a). Affective economies. *Social Text, 22*(2), 117–139. https://doi.org/10.1215/01642472-22-2_79-117

Ahmed, S. (2004b). *The cultural politics of emotion.* Edinburgh University Press.

Ahmed, S. (2010a). Foreword. In R. Ryan-Flood & R. Gill (Eds.), *Secrecy and silence in the research process: Feminist reflections* (pp. xvi–xxi). Routledge.

Ahmed, S. (2010b). *The promise of happiness.* Duke University Press.

Ahmed, S. (2014a). *Willful subjects.* Duke University Press.

Ahmed, S. (2014b). *The cultural politics of emotion* (2nd ed.). Edinburgh University Press.

Ahmed, S. (2017). *Living a feminist life.* Duke University Press.

Ahmed, S. (2019). *What's the use?: On the uses of use.* Duke University Press.

Alaimo, S. (2008). Trans-corporeal feminisms and the ethical space of nature. In S. Alaimo & S. J. Hekman (Eds.), *Material feminisms* (pp. 237–264). Indiana University Press.

Alaimo, S. (2016). *Exposed: Environmental politics and pleasures in posthuman times.* University of Minnesota Press.

Apunkt Schneider, F., & Friesinger, G. (2010). *Urban hacking as a practical and theoretical critique of public spaces.* transcript Verlag.

Arendt, H. (1968). *Men in dark times.* Harcourt, Brace and World.

Barad, K. (2007). *Meeting the universe halfway: Quantum physics and the entanglement of matter and meaning.* Duke University Press.

Barad, K. (2014). Diffracting diffraction: Cutting together-apart. *Parallax: Diffracted Worlds—Diffractive Readings: Onto-Epistemologies and the Critical Humanities, 20*(3), 168–187. https://doi.org/10.1080/13534645.2014.927623

137

Berlant, L. (2004). Introduction: Compassion (and withholding). In L. Berlant (Ed.), *Compassion: The culture and politics of an emotion* (pp. 1–14). Routledge.

Berlant, L. (2008). *The female complaint: The unfinished business of sentimentality in American culture.* Duke University Press.

Berlant, L. (2010). Cruel optimism. In M. Gregg & G. J. Seigworth (Eds.), *The affect theory reader* (pp. 93–117). Duke University Press.

Berlant, L. (2011). *Cruel optimism.* Duke University Press.

Berlant, L. (2016). The commons: Infrastructures for troubling times. *Environment and Planning D: Society and Space, 34*(3), 393–419. https://doi.org/10.1177/0263775816645989

Berlant, L., & Bojarska, K. (2019). The hundreds, observation, encounter, atmosphere, and world-making. *Journal of Visual Culture, 18*(3), 289–304. https://doi.org/10.1177/1470412919875404

Berlant, L., & Stewart, K. (2019). *The hundreds.* Duke University Press.

Bird Rose, D. B. (2004). *Reports from a wild country: Ethics for decolonisation.* University of New South Wales Press.

Bird Rose, D. (2015). Dialogue. In D. Bird Rose, R. Fincher & K. Gibson (Eds.), *Manifesto for living in the anthropocene* (pp. 127–131). Punctum Books. https://doi.org/10.21983/P3.0100.1.00

Bochner, A. P., & Ellis, C. (1992). Personal narrative as a social approach to interpersonal communication. *Communication Theory, 2*(2), 165–172. https://doi.org/10.1111/j.1468-2885.1992.tb00036.x

Bonenfant, Y. (2014). On sound and pleasure: Meditations on the human voice. *Sounding Out.* https://soundstudiesblog.com/2014/06/30/on-sound-and-pleasure-meditations-on-the-human-voice/#:~:text=%20On%20Sound%20and%20Pleasure%3A%20Meditations%20on%20the,Anglo-Sax-on%20standards%2C%20and%20her%20voice%20wa

Bullis, C. (2015). Retalking environmental discourses from a feminist perspective: The radical potential of ecofeminism. In J. G. Cantrill & C. L. Oravec (Eds.), *The symbolic earth: Discourse and our creation of the environment* (pp. 123–148). University Press of Kentucky.

Bunda, T. (2018). Seeing the Aboriginal sovereign warrior woman. *Lifted Brow, 40,* 4–5.

Bunda, T., & Phillips, L. G. (2023). Storying: The vitality of social movements. In L. G. Phillips & T. Bunda (Eds.), *Storying social movement/s* (Palgrave studies in movement across education, the arts and the social sciences). Palgrave.

Butler-Bowdon, E. (2001). Lawn and order: Aesthetics and architecture in Australian suburbia. *Studies in the History of Gardens & Designed Landscapes, 21*(2), 108–114. https://doi.org/10.1080/14601176.2001.10435240

Cadieux, K. V. (2013). Other women's gardens: Radical homemaking and public performance of the politics of feeding. In A. Hayes-Conroy & J. Hayes-Conroy (Eds.), *Doing nutrition differently: Critical approaches to diet and dietary intervention* (pp. 61–86). Ashgate.

Campbell, C. (2018). *The romantic ethic and the spirit of modern consumerism* (2nd ed.). Palgrave Macmillan.

Cantrell, C. (1994). Women and language in Susan Griffin's woman and nature: The roaring inside her. *Hypatia, 9*(3), 225–238. https://doi.org/10.1111/j.1527-2001.1994.tb00459

Capp, F. (2010). *My blood's country: In the footsteps of Judith Wright.* Allen & Unwin.

Ceccarelli, L. (1998). Polysemy: Multiple meanings in rhetorical criticism. *The Quarterly Journal of Speech, 84*(4), 395–415. https://doi.org/10.1080/00335639809384229

Chadha, M., Avila, A., & Gil de Zúñiga, H. (2012). Listening in: Building a profile of podcast users and analyzing their political participation. *Journal of Information Technology & Politics, 9*(4), 388–401. https://doi.org/10.1080/19331681.2012.717481

Cixous, H. (1976). The laugh of the medusa. *Signs, 1*(4), 875–893. https://doi.org/10.1086/493306

Cixous, H. (1993). *Three steps on the ladder of writing.* Columbia University Press.

Cixous, H. (2007). *Insister of Jacques Derrida.* Edinburgh University Press.

Cixous, H. (2008). *White ink: Interviews on sex, text and politics.* Acumen.

Cixous, H., & Calle-Gruber, H. (1997). *Rootprints: Memory and life writing.* Routledge.

Cixous, H., & Sellers, S. (1994). *The Hélène Cixous reader.* Routledge.

Compassion. (2021). *Merriam-Webster.* https://www.merriam-webster.com/dictionary/compassion

Copeland, S. (2018). A feminist materialisation of amplified voice: Queering identity and affect in The Heart. In D. Llinares, N. Fox, & R. Berry (Eds.), *Podcasting: New aural cultures and digital media* (pp. 209–225). Springer International Publishing AG.

Dean, J. (2009). *Democracy and other neoliberal fantasies communicative capitalism and left politics.* Duke University Press.

Dean, J. (2010). *Blog theory: Feedback and capture in circuits of drive.* Polity Press.

Dean, J. (2018). *Publicity's secret: How technoculture capitalizes on democracy.* Cornell University Press.

Diamond, I., & Orenstein, G. F. (1990). *Reweaving the world: The emergence of ecofeminism.* Sierra Club Books.

Didion, J. (1979). *The white album.* Simon & Schuster.

Dobson, A. S. (2015). *Postfeminist digital cultures femininity, social media, and self-representation.* Palgrave Macmillan.

Dobson, A. S., Carah, N., & Robards, B. (2018). Digital intimate publics and social media: Towards theorising public lives on private platforms. In A. S. Dobson, B. Robards, & N. Carah (Eds.), *Digital intimate publics and social media* (pp. 3–27). Palgrave Macmillan.

Fairlie, S. (2009). A short history of enclosure in Britain. *The Land.* https://www.thelandmagazine.org.uk/articles/short-history-enclosure-britain

Federici, S. (2019). *Re-enchanting the world: Feminism and the politics of the commons.* PM Press.

Francis, M., & Hester, R. (1990). *The meaning of gardens: Idea, place and action.* MIT Press.

Gaard, G. (1993). *Ecofeminism: Women, animals, nature*. Temple University Press.

Gibson, K. (2018). *Introduction: Food as urban commons and community economics*. University of Western Australia Publishing.

Gibson-Graham, J. K. (2006). *A postcapitalist politics*. University of Minnesota Press.

Glean. (2021). *Merriam Webster Dictionary* (Vol. 2020). https://www.merriam-webster.com/dictionary/glean

Glocer Fiorini, L. (2019). Deconstructing the feminine: Discourses, logics and power. Theoretico-clinical implications. *International Journal of Psychoanalysis, 100*(3), 593–603. https://doi.org/10.1080/00207578.2019.1590783

Griffin, S. (1978). *Woman and nature: The roaring inside her*. Harper & Row.

Griffin, S. (1982). The way of all ideology. *Signs, 7*(3), 641–660. https://doi.org/10.1086/493904

Griffin, S. (1993). Red shoes. In R.-E. B. Joeres & E. Mittman (Eds.), *The politics of the essay: Feminist perspectives* (pp. 1–11). Indiana University Press.

Griffin, S. (2015). *The eros of everyday life: Essays on ecology, gender and society*. Open Road Media.

Hamilton, J. M. (2019). The future of housework: The similarities and differences between making kin and making babies. *Australian Feminist Studies: What Do We Want? Feminist Environmental Humanities, 34*(102), 468–489. https://doi.org/10.1080/08164649.2019.1702874

Haraway, D. J. (2016). *Staying with the trouble: Making kin in the chthulucene*. Duke University Press.

Harvey, C. B. (1989). Some Irish women storytellers and reflections on the role of women in the storytelling tradition. *Western Folklore, 48*(2), 109–128. https://doi.org/10.2307/1499685

Hayes, S. (2010). *Radical homemakers: Reclaiming domesticity from a consumer culture*. Left to Write Press.

Haygood, D. M. (2007). A status report on podcast advertising. *Journal of Advertising Research, 47*(4), 518–523. https://doi.org/10.2501/S0021849907070535

Heller, C. (1993). For the love of nature: Ecology and the cult of the romantic. In G. Gaard (Ed.), *Ecofeminism: Women, animals, nature* (pp. 219–242). Temple University Press.

Hickel, J. (2020). Quantifying national responsibility for climate breakdown: An equality-based attribution approach for carbon dioxide emissions in excess of the planetary boundary. *The Lancet, 4*(9), e399–e404. https://doi.org/10.1016/S2542-5196(20)30196-0

Holmes, K. (1999). Gardens. *Journal of Australian Studies, 23*, 152–162. https://doi.org/10.1080/14443059909387485

Holmes, K. (2011). *Between the leaves: Stories of Australian women, writing and gardens*. UWA Publishing.

Honan, E., & Bright, D. (2016). Writing a thesis differently. *International Journal of Qualitative Studies in Education, 29*(5), 731–743. https://doi.org/10.1080/09518398.2016.1145280

hooks, b. (1997). *Wounds of passion: A writing life*. Henry Holt.

hooks, b. (1999). *Remembered rapture: The writer at work*. Henry Holt.

hooks, b. (2001). *All about love*. Harper Collins.

hooks, b. (2003). *Communion: The female search for love*. Perennial.

Hutchison, E. (2010). Unsettling stories: Jeanette Winterson and the cultivation of political contingency. *Global Society, 24*(3), 351–368.

Hutchison, E. (2016). *Affective communities in world politics: Collective emotions after trauma*. Cambridge University Press.

Jackson, M. (2002). *The politics of storytelling: Violence, transgression, and intersubjectivity*. Museum Tusculanum Press.

Jackson, M. (2013). *The politics of storytelling: Variations on a theme by Hannah Arendt* (2nd ed.). Museum Musculanum Press.

Jacobs, R. N. (2002). The Narrative Integration of Personal and Collective Identity in Social Movements. In *Narrative Impact* (1st ed., pp. 205–228). Routledge. https://doi.org/10.4324/9781410606648-11

Johnson, P. A. (2008). The howl that could not be silenced: The rise of queer radio. In M. C. Keith (Ed.), *Radio cultures: The sound medium in American life* (pp. 95–112). Peter Lang.

Johnston, J. (2006). Who cares about the commons? In M. A. Gismondi, J. Goodman, & J. Johnston (Eds.), *Nature's revenge: Reclaiming sustainability in an age of corporate globalization* (pp. 39–72). Broadview Press.

Johnston, J. (2008). Counterhegemony or bourgeois piggery? food politics and the case of foodshare. In Middendorf, G., & Wright, W (Eds.). *The fight over food : producers, consumers, and activists challenge the global food system*. Pennsylvania State University Press.

Johnston, J., & Cairns, K. (2012). Eating for Change. In R. Mukherjee, & S. Banet-Weiser (Eds.), *Commodity activism: Cultural resistance in neoliberal times*. New York University Press.

Knowles, C., & Sweetman, P. (2004). Introduction. In C. Knowles & P. Sweetman (Eds.), *Picturing the social landscape visual methods and the sociological imagination* (pp. 1–17). Routledge.

Kuntsman, A. (2012). Introduction: Affective fabrics of digital cultures. In A. Karatzogianni & A. Kuntsman (Eds.), *Digital cultures and the politics of emotion* (pp. 1–17). Palgrave Macmillan.

Kwaymullina, A. (2018). You are on Indigenous land: Ecofeminism, Indigenous peoples and land justice. In L. Stevens, P. Tait, & D. Varney (Eds.), *Feminist ecologies: Changing environments in the anthropocene* (pp. 193–208). Palgrave Macmillan.

Langellier, K. M. (1989). Personal narratives: Perspectives on theory and research. *Text and Performance Quarterly, 9*(4), 243–276. https://doi.org/10.1080/10462938909365938

Langellier, K. M., & Peterson, E. E. (2011). *Storytelling in daily life: Performing narrative*. Temple University Press.

Le Guin, U. K. (1989). *Dancing at the edge of the world: Thoughts on words, women, places*. Grove Press.

Le Guin, U. K. (2004). *The wave in the mind: Talks and essays on the writer, the reader, and the imagination* (1st ed.). Shambhala.

Lewis, T. (2012). "There grows the neighbourhood": Green citizenship, creativity and life politics on eco-TV. *International Journal of Cultural Studies, 15*(3), 315–326.

Lipton, B., & Mackinlay, E. (2017). *We only talk feminist here: Feminist academics, voice and agency in the neoliberal university*. Springer International Publishing.

Llinares, D., Fox, N., & Berry, R. (2018). Introduction: Podcasting and podcasts-parameters of a new aural culture. In D. Llinares, N. Fox, & R. Berry (Eds.), *Podcasting: New aural cultures and digital media* (pp. 1–13). Springer International Publishing AG.

Lucashenko, M. (2003). Not quite white in the head. *Griffith Review, 2*, 7–15.

MacGregor, S. (2004). From care to citizenship: Calling ecofeminism back to politics. *Ethics & the Environment, 9*(1), 56–84. https://doi.org/10.2979/ETE.2004.9.1.56

MacGregor, S. (2006). *Beyond mothering earth: Ecological citizenship and the politics of care*. UBC Press.

Mackinlay, E. (2015). Making an appearance on the shelves of the room we call research: Autoethnography-as-storyline-as-interpretation in education. In P. Smeyers (Ed.), *International handbook of interpretation in educational research* (pp. 1437–1456). Springer.

Mackinlay, E. (2016a). The heartlines in your hand. In E. Emerald, R. E. Rinehart, & A. Garcia (Eds.), *Global south ethnographies: Minding the senses* (pp. 153–165). Sense Publishers.

Mackinlay, E. (2016b). *Teaching and learning like a feminist storying our experiences in higher education*. Sense Publishers.

Mackinlay, E. (2019). *Critical writing for embodied approaches: Autoethnography, feminism and decoloniality*. Palgrave Macmillan.

Mackinlay, E. (2022). *Writing feminist autoethnolgraphy: In love with theory, words, and the language of women writers*. Routledge.

Mackinlay, E., & Bartleet, B.-L. (2012). Exploring the potential of sisterhood and personal relationships as the foundations of musicological and ethnographic fieldwork. *Qualitative Research Journal, 12*(1), 75–87. https://doi.org/10.1108/14439881211222741

Mackinlay, E., & Madden, K. (Eds.). (2024). *Departing radically in academic writing: Alternative approaches to writing methods in qualitative research*. Routledge.

Maniates, M. F. (2001). Individualization: Plant a Tree, Buy a Bike, Save the World? *Global Environmental Politics, 1*(3), 31–52. https://doi.org/10.1162/152638001316881395

Martínez Luna, S. (2019). Still images? Materiality and mobility in digital visual culture. *Third Text, 33*(1), 43–57. https://doi.org/10.1080/09528822.2018.1546484

McDonagh, B. (2013). Making and breaking property: Negotiating enclosure and common rights in sixteenth-century England. *History Workshop Journal, 76*(1), 32–56. https://doi.org/10.1093/hwj/dbs054

McDonagh, B. (2019a). Disobedient objects: Material readings of enclosure protest in sixteenth-century England. *Journal of Medieval History, 45*(2), 254–275. https://doi.org/10.1080/03044181.2019.1593629

McDonagh, B. (2019b). *Gendering protest and the commons.* Paper presented at the Proceedings of the International Conference of Historical Geographers, Warsaw, Poland.

McDonagh, B., & Daniels, S. (2012). Enclosure stories: Narratives from Northamptonshire. *Cultural Geographies, 19*(1), 107–121. https://doi.org/10.1177/1474474011427361

McDonagh, B., & Griffin, C. J. (2016). Occupy! Historical geographies of property, protest and the commons, 1500–1850. *Journal of Historical Geography, 53*, 1–10. https://doi.org/10.1016/j.jhg.2016.03.002

McWhorter, L. (1999). *Bodies and pleasures: Foucault and the politics of sexual normalization.* Indiana University Press.

Merchant, C. (2010). *Ecological revolutions: Nature, gender, and science in New England* (2nd ed.). The University of North Carolina Press.

Mickelburgh, R. (2020). Compassion in the garden: Radical homemakers or just more women's work? *Emotions: History, Culture, Society, 4*(1), 146–166. https://doi.org/10.1163/2208522X-02010092

Mickelburgh, R. (2020b). Unearthing women's activism. *Australian Garden History Journal, 32*(2), 8–11.

Mickelburgh, R. (2023). Writing strange letters in the garden, with Love and Fury. *Swamphen: A Journal of Cultural Ecology, 9*, 1–6.

Mies, M. (1998). *Patriarchy and accumulation on a world scale: Women in the international division of labour* (New ed.). Zed Books.

Mies, M. (2014a). Liberating the consumer. In V. Shiva, M. Mies, & A. Salleh (Eds.), *Ecofeminism* (pp. 251–263). Zed Books.

Mies, M. (2014b). No commons without a community. *Community Development Journal, 49*(SI), i106–i117. https://doi.org/10.1093/cdj/bsu007

Mies, M., & Bennholdt-Thomsen, V. (1999). *The subsistence perspective: Beyond the globalised economy.* Zed Books.

Mies, M., & Bennholdt-Thomsen, V. (2001). Defending, reclaiming and reinventing the commons. *Canadian Journal of Development Studies, 22*(4), 997–1023. https://doi.org/10.1080/02255189.2001.9669952

Mies, M., & Shiva, V. (2014). *Ecofeminism.* Zed Books.

Moreton-Robinson, A. (2000). *Talkin' up to the white woman: Aboriginal women and feminism.* University of Queensland Press.

Moreton-Robinson, A. (2015). *The white possessive: Property, power, and indigenous sovereignty.* University of Minnesota Press.

Mortimer-Sandilands, C. (2008). Landscape, memory, and forgetting: Thinking through (my mother's) body and place. In S. J. Hekman & S. Alaimo (Eds.), *Material feminisms* (pp. 265–288). Indiana University Press.

Mulligan, M., & Hill, S. (2001). *Ecological pioneers: A social history of Australian ecological thought and action*. Cambridge University Press.

Padilla Carroll, V. (2016). The radical possibilities of new (feminist, environmentalist) domesticity: Housewifery as an altermodernity project. *Interdisciplinary Studies in Literature and Environment, 23*(1), 51–70. https://doi.org/10.1093/isle/isw013

Phillips, L. G., & Bunda, T. (2018). *Research through, with and as storying.* Routledge.

Phillips, M. (2016). Embodied care and planet earth: Ecofeminism, maternalism and postmaternalism. *Australian Feminist Studies, 31*(90), 468–485. https://doi.org/10.1080/08164649.2016.1278153

Pip Magazine. (n.d.). *Pip Podcasts.* https://pipmagazine.com.au/category/podcasts/

Planthunter. (n.d.). *The Planthunter.* https://theplanthunter.com.au/

Plumwood, V. (1993). *Feminism and the mastery of nature.* Routledge.

Plumwood, V. (1994). The ecopolitics debate and the politics of nature. In K. Warren (Ed.), *Ecological feminism* (pp. 64–87). Routledge.

Plumwood, V. (1999). Ecological ethics from rights to recognition. In N. P. Low (Ed.), *Global ethics and environment* (pp. 202–226). Routledge.

Plumwood, V. (2002). *Environmental culture: The ecological crisis of reason.* Routledge.

Plumwood, V. (2007). Journey to the heart of stone. *Nature, Culture and Literature, 5,* 17.

Plumwood, V. (2009). Nature in the active voice. *Australian Humanities Review, 46,* 113–129. https://doi.org/10.22459/AHR.46.2009.10

Plumwood, V. (2018). Ecofeminist analysis and the culture of ecological denial. In L. Stevens, P. Tait, & D. Varney (Eds.), *Feminist ecologies: Changing environments in the anthropocene* (pp. 97–112). Palgrave Macmillan.

Poletti, A. (2020). *Stories of the self: Life writing after the book.* New York University Press.

Polletta, F. (2006). *It was like a fever: Storytelling in protest and politics.* University of Chicago Press.

Reid, G. (2017a, July 4). River garden diaries: Truth, Thoreau and design dilemmas. *The Planthunter.* https://theplanthunter.com.au/gardens/river-garden-diaries-truth-thoreau-design/

Reid, G. (2017b, October 17). River garden diaries: Tree songs and a sense of place. *The Planthunter.* https://theplanthunter.com.au/gardens/river-garden-diaries-tree-songs-sense-place/

Reid, G. (2018a, March 20). River garden diaries: Neighbourly relations. *The Planthunter.* https://theplanthunter.com.au/gardens/river-garden-diaries-neighbourly-relations/

Reid, G. (2018b, July 18). River garden diaries: A book in a boatshed and pondering place. *The Planthunter.* https://theplanthunter.com.au/culture/river-diaries-book-boatshed-pondering-place/

Reid, G. (2018c, August 21). River garden diaries: Composting jellyfish and other meanderings. *The Planthunter*. https://theplanthunter.com.au/gardens/river-garden-diaries-composting-jellyfish/

Reid, G. (2018d, October 17). The way of the gardener. *The Planthunter*. https://theplanthunter.com.au/culture/the-way-of-the-gardener/

Reid, G. (2018e, October 24). River garden diaries: And then the snake (and the rain) came. *The Planthunter*. https://theplanthunter.com.au/gardens/river-garden-diaries-snake-rain-came/

Reid, G. (2019a, June 11). Say my name: On speaking the Indigenous names of plants. *The Planthunter*. https://theplanthunter.com.au/culture/say-name-speaking-indigenous-names-plants/

Reid, G. (2019b, August 7). Audacious gardening: On daring to care. *The Planthunter*. https://theplanthunter.com.au/gardens/audacious-gardening-daring-care/

Richardson, L. (1997). *Fields of play: Constructing an academic life*. Rutgers University Press.

Richardson, L. (2017). Writing: A method of inquiry. In N. K. Denzin & Y. S. Lincoln (Eds.), *The Sage handbook of qualitative research* (5th ed., pp. 1410–1444). Sage.

Riessman, C. (2008). *Narrative methods for the human sciences*. Sage.

Robbins, P. (2007). *Lawn people: How grasses, weeds, and chemicals make us who we are*. Temple University Press.

Robbins, P., & Sharp, J. T. (2003). Producing and consuming chemicals: The moral economy of the American Lawn. *Economic Geography, 79*(4), 425–451. https://doi.org/10.1111/j.1944-8287.2003.tb00222.x

Rose, G. (2014). On the relation between "visual research methods" and contemporary visual culture. *The Sociological Review, 62*(1), 24–46. https://doi.org/1 0.1111/1467-954X.12109

Rosenfeldt, R. (Host) (2017a). #8 Women as changemakers [Audio podcast episode] Pip Podcast. *Pip Magazine*. https://www.pipmagazine.com.au/podcasts/pip-podcast-8-women-change-makers/

Rosenfeldt, R. (Host) (2017b). # 3: What is permaculture? with Hannah Moloney [Audio podcast episode] Pip Podcast. *Pip Magazine*. https://www.pipmagazine.com.au/podcasts/what-is-permaculture-hannah-moloney/

Rosenfeldt, R. (Host) (2018a). #10 Mariam Issa [Audio podcast episode]. Pip Podcast. *Pip Magazine*. https://www.pipmagazine.com.au/podcasts/pip-permaculture-podcast-10-mariam-issa/

Rosenfeldt, R. (Host) (2018b). #11 Jodie Vennitti [Audio podcast episode]. Pip Podcast. *Pip Magazine*. https://pipmagazine.com.au/news/pip-permaculture-podcast-11-jodie-vennitti/

Rubin, L. B. (1985). *Just friends: The role of friendship in our lives*. Harper & Row.

Sandberg, L. A., & Foster, J. (2005). Challenging lawn and order: Environmental discourse and lawn care reform in Canada. *Environmental Politics, 14*(4), 478–494. https://doi.org/10.1080/09644010500175692

Sandilands, C. (1999). *The good-natured feminist: Ecofeminism and the quest for democracy.* University of Minnesota Press.

Singer, N. R. (2020). Toward Intersectional Ecofeminist Communication Studies. *Communication Theory, 30*(3), 268–289. https://doi.org/10.1093/ct/qtz023

Singer, N. R. (2021). Toward Intersectional Ecofeminist Communication Studies (vol 30, pg 268, 2020). *Communication Theory, 31*(4), 1022–1022. https://doi.org/10.1093/ct/qtaa020

Smith, N., & Walters, P. (2018). Desire lines and defensive architecture in modern urban environments. *Urban Studies, 55*(13), 2980–2995. https://doi.org/10.1177/0042098017732690

Smith, T. (2014). *The myth of green marketing: Tending our gods at the edge of apocalypse.* University of Toronto Press.

Somerville, M. (2004). *Wildflowering: The life and places of Kathleen McArthur.* University of Queensland Press.

Soper, K. (2007a). The other pleasures of post-consumerism. *Soundings: A Journal of Politics and Culture, 35,* 31–40. https://doi.org/10.3898/136266207820465930

Soper, K. (2007b). Re-thinking the good life: The citizenship dimension of consumer disaffection with consumerism. *Journal of Consumer Culture, 7*(2), 205–229. https://doi.org/10.1177/1469540507077681

Spelman, E. V. (1997). *Fruits of sorrow: Framing our attention to suffering.* Beacon Press.

Spretnak, C. (1990). Ecofeminism: Our roots and flowering. In I. Diamond & G. F. Orenstein (Eds.), *Reweaving the world: The emergence of ecofeminism* (pp. 3–14). Sierra Club Books.

St. Pierre, E. A. (2018). Writing post qualitative inquiry. *Qualitative Inquiry, 24*(9), 603–608. https://doi.org/10.1177/1077800417734567

Stephens, J. (2012). *Confronting postmaternal thinking: Feminism, memory, and care.* Columbia University Press.

Stevens, L. (2018). From the female eunuch to White beech: Germaine Greer and ecological feminism. In L. Stevens, P. Tait, & D. Varney (Eds.), *Feminist ecologies: Changing environments in the anthropocene* (pp. 115–133). Palgrave Macmillan.

Stevens, L., Tait, P., & Varney, D. (2018). Introduction: "Street-fighters and philosophers": Traversing ecofeminisms. In L. Stevens, P. Tait, & D. Varney (Eds.), *Feminist ecologies: Changing environments in the anthropocene* (pp. 1–22). Palgrave Macmillan.

Stevens, S. (2009). The official rhetoric of permaculture: Motivating behaviour change through environmental communication. *Australian Journal of Communication, 36*(2), 73–91.

Stewart, K. (2007). *Ordinary affects.* Duke University Press.

Stewart, K. (2010). Afterword: Worlding refrains. In M. Gregg & G. J. Seigworth (Eds.), *The affect theory reader* (pp. 239–253). Duke University Press.

Tiffe, R., & Hoffmann, M. (2017). Taking up sonic space: Feminized vocality and podcasting as resistance. *Feminist Media Studies, 17*(1), 115–118. https://doi.org/10.1080/14680777.2017.1261464

Tuck, E., & Yang, K. W. (2012). Decolonization is not a metaphor. *Decolonization: Indigeneity, Education and Society, 1*(1), 1–40.

Urban Food Street. (2016a, 3 January). *Happy detouring!* [Image attached]. Facebook. https://www.facebook.com/urbanfoodstreet/photos/a.39562 5327284340/444088652438007/

Urban Food Street. (2016b, 10 February). *Roadways as common ground.* [Image attached]. Facebook. https://www.facebook.com/urbanfoodstreet/photos/487343611445844

Urban Food Street. (2016c, 13 March). *Children help making gardens.* [Image attached]. Facebook. https://www.facebook.com/urbanfoodstreet/photos/496379413875597

Urban Food Street. (2016d, 25 April). *Exposed soil.* [Image attached]. Facebook. https://www.facebook.com/urbanfoodstreet/photos/512796875567184

Urban Food Street. (2016e, 12 June). *Lawn verge undergoes change-of-use.* [Image attached]. Facebook. https://www.facebook.com/urbanfoodstreet/photos/528475027332702

Urban Food Street. (2016f, 29 June). *Crops replace lawn on the verge.* [Image attached]. Facebook. https://www.facebook.com/urbanfoodstreet/photos/534042480109290

Urban Food Street. (2016g, 24 July). *Smiling faces grow in the Urban Food Street garden.* [Image attached]. Facebook. https://www.facebook.com/urbanfood-street/photos/542765285903676

Urban Food Street. (2016h, 29 September). *Human traffic.* [Image attached]. Facebook. https://www.facebook.com/urbanfoodstreet/photos/5677955 60067315

Urban Food Street. (2016i, 9 October). *Creating verge gardens.* [Image attached]. Facebook. https://www.facebook.com/urbanfoodstreet/photos/57170 2993009905

Urban Food Street. (2016j, 9 October). *Gardening tools: Mundane or disruptive?* [Image attached]. Facebook. https://www.facebook.com/urbanfoodstreet/photos/571714109675460

Urban Food Street. (2016k, 9 October). *Residents digging up lawn verge.* [Image attached]. Facebook. https://www.facebook.com/urbanfoodstreet/photos/571703203009884

Urban Food Street. (2016l, 27 October). *Smiling faces grow in the Urban Food Street garden.* [Image attached]. Facebook. https://www.facebook.com/urbanfoodstreet/photos/579563872223817

Urban Food Street. (2016m, 29 November). *Children play after Christmas celebrations.* [Image attached]. Facebook. https://www.facebook.com/urban-foodstreet/photos/593471497499721

Urban Food Street. (2016n, 7 August). *A joyful harvest*. [Image attached]. Facebook. https://www.facebook.com/urbanfoodstreet/photos/pcb.54788 3598725178/547883108725227/

Urban Food Street. (2016o, 8 June). *Gardening tools recall another era*. [Image attached]. Facebook. https://www.facebook.com/urbanfoodstreet/photos/ 1307231679457029

Urban Food Street. (n.d.). *Facebook: Urban Food Street*. https://www.facebook. com/urbanfoodstreet

van Dijck, J. (2004). Mediated memories: Personal cultural memory as object of cultural analysis. *Continuum: Journal of Media and Cultural Studies, 18*(2), 261–277. https://doi.org/10.1080/1030431042000215040

van Dijck, J. (2007). *Mediated memories in the digital age*. Stanford University Press.

Walker, M. B. (2016). *Slow philosophy: Reading against the Institution*. Bloomsbury Academic.

Wall Kimmerer, R. (2013). *Braiding sweetgrass*. Penguin Books.

Warren, K. (1994). Introduction. In K. Warren (Ed.), *Ecological feminism* (pp. 1–7). Routledge.

Warren, K. (1996). The power and the promise of ecological feminism. In K. Warren (Ed.), *Ecological feminist philosophies* (Vol. 12, pp. 125–146). Indiana University Press.

White, H. V. (1987). *The content of the form: Narrative discourse and historical representation*. Johns Hopkins University Press.

Whitebrook, M. (2001). *Identity, narrative, and politics*. Routledge.

Whyte, N. (2011). Custodians of memory: Women and custom in Rural England c. 1550–1700. *Cultural and Social History, 8*(2), 153–173. https://doi.org/1 0.2752/147800411X12949180694263

Whyte, N. (2015). Senses of place, senses of time: Landscape history from a British perspective. *Landscape Research, 40*(8), 1–14. https://doi.org/10.108 0/01426397.2015.1074987

Wichterich, C. (2015). Contesting green growth, connecting care, commons and enough. In W. Harcourt & I. L. Nelson (Eds.), *Practising feminist political "ecologies": Moving beyond the "green economy"* (pp. 67–100). Zed Books.

Winskell, K., & Enger, D. (2014). Storytelling for social change. In *The handbook of development and social change* (pp. 189–206). John Wiley & Sons.

Winterson, J. (2012). *Why be happy when you could be normal?* Grove.

Woolf, V. (1929/1977). *A room of one's own*. Granada.

Woolf, V. (1952a). *The leaning tower* (Uniform ed.). Hogarth Press.

Woolf, V. (1952b). The moment: Summer's night. In V. Woolf (Ed.), *The moment and other essays* (Uniform ed., pp. 9–13). Hogarth Press.

Wright, J., Clarke, P., McKinney, M., & National Library of Australia. (2006). *With love and fury: selected letters of Judith Wright*. National Library of Australia.

Wright, J., & Snibson, D. (2014). *The coral battleground*. Spinifex.

Index[1]

[1] Note: Page numbers followed by 'n' refer to notes.

© The Author(s), under exclusive license to Springer Nature Switzerland AG 2024
R. Mickelburgh, *The Ecofeminist Storyteller*,
https://doi.org/10.1007/978-3-031-59242-3